The Future of Music

Guerino Mazzola • Jason Noer • Yan Pang •
Shuhui Yao • Jay Afrisando •
Christopher Rochester • William Neace

The Future of Music

Towards a Computational Musical Theory of Everything

Springer

Guerino Mazzola
School of Music
University of Minnesota
Minneapolis, USA

Jason Noer
Department of Theatre Arts & Dance
University of Minnesota
Minneapolis, USA

Yan Pang
Department of Theatre Arts & Dance
University of Minnesota
Minneapolis, USA

Shuhui Yao
School of Music
University of Minnesota
Minneapolis, USA

Jay Afrisando
School of Music
University of Minnesota
Minneapolis, USA

Christopher Rochester
School of Music
University of Minnesota
Minneapolis, USA

William Neace
School of Music
University of Minnesota
Minneapolis, USA

ISBN 978-3-030-39708-1 ISBN 978-3-030-39709-8 (eBook)
https://doi.org/10.1007/978-3-030-39709-8

This Springer imprint is published by the registered company Springer Nature Switzerland AG.
The registered company address is: Gewerbestrasse 11, 6330 Cham, Switzerland

Dedicated to Ida Mazzola

Photo by Guerino Mazzola

Preface

The idea for this book came from a reading of the trilogy *The Three-Body Problem, The Dark Forest, Death's End* by the famous science fiction author Cixin Liu, which was recommended to me by my student Shuhui Yao. It was less the concrete narrative of these books—spread over the entire universe in time and space—than the very category of "science fiction" which was troubling. Science fiction deals with future and more or less fictitious perspectives of the world's physical reality, together with narratives that inhabit those fictions.

It prompted the authors to ask themselves as creators, theorists, and performers of music and dance whether there was something like "music fiction," a narrative in a *musical* fictitious future.

While this book is inspired by a science fiction, this book's proposed conceptualization of music fiction is not about fiction of a musical future. Instead, it is a historical and theoretical investigation with the purpose of predicting the future of music.

We started our inquiry by thinking about how music—instead of science—could look and feel in 50 years. As Guerino Mazzola was then teaching his university course "The Mathematical Design of Future Music," his next step was to envisage a project about the possible realms of future music.

In the Fall 2018, Mazzola therefore offered a graduate course on writing a book following these ideas. The precise goal of this project was to envisage a book that would be the first step towards what we have come to call a "Computational Music Theory of Everything" (ComMuTE). This title was borrowed from what in physics is now the Big Science ideal for finding a theory that explains everything. This "Theory of Everything" (ToE) would unify the four fundamental forces: electromagnetic, weak, strong, and gravitational. The unification of the electromagnetic and weak to the electro-weak being the first big step towards ToE.

We, the collaborators, who took part in this collective effort, represent diverse experiences and contribute expertise in specialities including composition, improvization, and dance. This collaborative effort offers not only new directions and perspectives that arise from their routine work with creativity that includes new technologies, but also new interpretations of the social and aesthetic role of music in its multicultural global variety.

This book is the result of our collaboration that was extended to the Spring 2019 semester and offers a breathtaking spectrum of new vectors into the future of music. In the general introduction, Chap. 1, a perspective of future developments will be drawn on the basis of the book's chapters.

We are pleased to acknowledge the strong support for writing such a demanding treatise from Springer's science editor Thomas Hempfling.

Minneapolis, USA — Guerino Mazzola
Minneapolis, USA — Jason Noer
Minneapolis, USA — Yan Pang
Minneapolis, USA — Shuhui Yao
Minneapolis, USA — Jay Afrisando
Minneapolis, USA — Christopher Rochester
Minneapolis, USA — William Neace
May 2019

From left to right: William Neace, Jay Afrisando, Jason Noer, Yan Pang, Guerino Mazzola, Shuhui Yao, Christopher Rochester. Image reprinted with kind permission from the Authors

Contents

Part I
Introduction

Chapter 1
General Introduction

1.1 The Collaborative Authors

To write a book about the future of music, the strategies in composition and technology, is a delicate enterprise. Beyond a display of theoretical, technological, and practical tools, one has to sketch the cultural and sociological perspectives in a world that changes at a very high pace.

What type of authorship should one set up to this end? There are two polar forces here: one the one hand, the maximal experience in the development of theoretical, technological, and practical components should be addressed. This role could be taken by author Guerino Mazzola, who has worked as a music theorist, music software programmer, and jazz musician for over four decades. But on the other hand, the future of music must also be investigated by those who will have to shape musical creativity in the near future: young composers, musicians, dancers, or students of these fields. The perspectives of their own vital future directions are as important as those given by long experience. For this reason, the present authors were chosen to provide the reader with a balanced account of experience and prospects. Jason "J-Sun" Noer is a disciplinary head of the Hip Hop dance track at the University of Minnesota. Yan Pang is a professional composer and dance faculty at the University of Minnesota, opening the multicultural perspective between Western and Chinese traditions. Shuhui Yao is a smart master's student, Jay Afrisando, William Neace, and Christopher Rochester are very engaged PhD students of composition.

G. Mazzola et al., *The Future of Music*,
https://doi.org/10.1007/978-3-030-39709-8_1

1.2 Not State of the Art, but Milestones to the Future

The dominant vector of this book is not a description of the state of the art, but a set of strong directives where to go, and where not to go. This is its provocative character, imagining and designing the unknown future.

1.3 Outdated Principles

The book's directive vectors should comprise a critique the following outdated principles/concepts. Let us state these principles as components of a traditional ideal creative musician. Observe that in the traditional understanding, a musician is always masculine because of the outdated ideology that women cannot create music. Here is a leading general ideological statement, where the outdated principles/concepts are included:

*He is the **divinely inspired genius** who expresses his deep **emotions** of (biological) **survival** for the distinguished* ***artist** in **works** of **individual excellence**.*

Here are principles which would replace those outdated ones:

1. **creative engineering** replacing **genius**, i.e., the systematic investigation of creative strategies built upon creativity theory
2. **free flow of information** replacing **divine inspiration**, i.e., not being closed, but sensitive to all input, but also filtering out invalid alternatives
3. **abolish split inside/outside** replacing **individualism**, i.e., the "individual" was the non-split, the box that defined inside and outside. This is outdated, we are a kind of Moebius space
4. **counterworlds** replacing **biological survival motivation**, i.e., counterworlds are about survival of mental states, not biological entities. This is much more important than a biological function
5. **instant mirroring of poiesis with aesthesis** replacing **dualism of artist vs. audience**, i.e., instant switch of roles, the linear model is replace by an "infinitesimal" transfer poiesis <—> aesthesis, i.e., the passionate activity: passive and active at the same time
6. **gestural mediation** replacing **dominance of emotions**, i.e., emotions are too limited, also too disembodied
7. **continuous interaction with knowledge** replacing **work concept**, i.e., works are static, like dead bodies, versions in time replace static works
8. **basic vital function** replacing **artistic vs. business**, i.e., breathing the world must be free of charges.

And here is the characterization of a person/musician who would embody these eight points:

- *(What?) He/She uses the free flow of information for creative engineering to elaborate the gestural dynamics of sounding counterworlds*

- *(How?) in an incessant interaction with knowledge on both sides of the cognitive Moebius space*
- *(What for?) to breathe his/her life in its mirror of passionate activity.*

1.3.1 A Total Reengineering of Music

There is something larger that underlies all the music principles, but which we cannot name yet. The idea of a *Computational Music Theory of Everything* (ComMute) is pointing in the right direction, but it is quite abstract and socially and economically vague. The applied science of engineering is a realization in the case of physics. But do we have an analogous situation of "musical engineering"? Such questions are paramount when talking about the future of music.

The traditional situation, which has defined the present state as described above, looks like asking for a total reengineering of the concept and reality of music, theoretically, technologically, socially, and financially, see also [148]. The reason for this impression is given by the comparison to physics. It may be a strange comparison, but it definitely opens perspectives that were hidden to this date.

1.4 The Book's Architecture of Future Musical Perspectives

In this section we present the book's parts and chapters in their role as vectors to future developments.

1.4.1 Part I: Introduction

This part gives a general orientation about the music's general ontology and role.

1.4.1.1 Chapter 1: General Introduction

This chapter!

1.4.1.2 Chapter 2: Ontology and Oniontology

This short chapter introduces the global architecture of ontology of music, which this book is going to use extensively. It provides the reader with the generic modes of being of music.

1.4.1.3 Chapter 3: The Basic Functions of Music

In this chapter we will focus on the basic functions of music and divide its focus into two distinguished approaches, theoretical and empirical. In the theoretical approach we will deduce the basic function of music based on theoretical frameworks based on evolutionary and non-evolutionary approaches. Within the empirical functions we will discuss the role of music in daily life, as well as discover potential fundamental dimensions implied by the multiple functions of music. Then we discuss how music acts on the human brain like a drug through the hippocampal gate function. This gives us an overview of all potential functions, for past, present, and future times.

1.4.1.4 Chapter 4: Historicity in Music

Music appears also as a function of historical determinants. How is music determined by such factors? What could be New Music versus Old Music? We argue against such fashionable but superficial terms. Music shares a deep connection to its temporal coordinates.

1.4.1.5 Chapter 5: Only One Restriction: Quality

Quality is a concept that greatly varies based on the predilection we develop throughout our lifetimes. In order for us to create a hypothesis on the trajectory of quality in future music, we need to establish the criterion in which we can attempt to objectively discern the value in all music. We discuss the methodology we have created to limit the innate prejudices developed throughout our lifetimes and aim to create an equitable means of categorization in which we value quality in conception through a combination of multiple facets.

1.4.2 Part II: Technological Tools

This part gives an overview of the technological perspectives of the musical development.

1.4.2.1 Chapter 6: Software Tools and Hardware Options

With the advent of recording and playback, the role of tools in the process of composing continues to grow its importance. Modern composing tools are not limited to the instant feedback of sound, but also have a great boost in visual aid. As for whether the future hardware and software can go beyond the physical level, for example, providing composers with psychological analysis etc., it is still a remaining question. But today's tools have been able to provide strong support in composition, analysis, (improvisation) performance, notation, and education. We discuss some of these tools together with their impact on musical creativity. Hardware for music is increasingly powerful for storage and processing purposes. We discuss this vector for its role in present musical productivity and its future.

In Sect. 6.4, an application of software for rhythm analysis (MetroRubette) to Brahms's Sonata op. 1 is described. This analysis is coauthored by Guerino Mazzola, Ruhan Alpaydin, and Bill Heinze.

1.4.2.2 Chapter 7: New Concepts of Musical Instruments

As we create a means to cogitate the future of music composition, we need to consider the relationship between composer and the medium in which music is performed. The concept of the development of musical instruments finds its foundation in the need for musicians to create sounds in ways that require more than one's own physical body. Currently we organize musical instruments in a system based on the production modality of sound. We distinguish acoustic instruments, which generate sounds in a mechanical way, from electroacoustic instruments which create sounds by electromagnetic devices. With the groundwork of instruments being steadily refined, the focus of *development* of musical instruments can generally be accomplished by two strategies. The first is the expansive realization of an instrument, i.e., the improvement of the versatility of an existing instrument. The second is the creative realization of a new instrument, which has been the primary focus of development for the majority of music.

1.4.2.3 Chapter 8: Musical Distribution Channels

Distribution of music is an essential component of its social relevance. It can more and more be a core factor of the music's socially relevant shape, in particular when given the direction of a quasi-simultaneous network resonance. This development will be essential for future musical productivity.

1.4.2.4 Chapter 9: Big Science in Music

In physics, Big Science is defined by a theory (models and theorems), an experimental paradigm, extensive laboratories, international collaboration, and political acceptance. In this chapter, we discuss such an option for the science of music.

1.4.3 Part III: Mathematical Concepts

This part presents some basic mathematical concept architectures for a future music.

1.4.3.1 Chapter 10: Mathematical Music Theory

Mathematical Music Theory offers a new view on musical conceptualization, in particular the power of generalization of models and theorems for musical phenomena. We stress the creative power of this approach, as opposed to traditional conservative music theories, and also as a tool for the music's future constructions.

1.4.3.2 Chapter 11: Serialism: Failure of New Concepts Without Musical Impact

We discuss the serial approach, in particular exemplified on Boulez's *Structures pour deux pianos*. We prove the failure of this approach in terms of musical parameters: communication, understanding, and creative power. This chapter should act as a caveat in front of mislead conceptualizations.

1.4.3.3 Chapter 12: Mazzola's Sonata Construction: A Technical Approach and Its Limits

This chapter gives a prototypical example of mathematical methods that should play a role in musical composition of the future. Mazzola's sonata is an explicit experiment to understand the mechanisms of sonata construction and its creative limitations, starting from the analysis of Beethoven's *Hammerklavier Sonata Op. 106* to compose a new sonata. This is an example of Boulez's creative analysis. We then critically review the Boulezian idea and the new composition.

1.4.3.4 Chapter 13: Imaginary Time

Imaginary time is a creative extension of musical time concepts, including a synthesis of Descartes' dual theory of mind versus body and its consequences for the understanding of consciousness and musical reality. This could provide us with a new understanding of time that should be important for the future of musical ontology.

1.4.3.5 Chapter 14: Mathematical Gesture Theory

The theory of gestures in music is a new approach to musical shaping forces. We discuss its potential (pre)semiotic impact on the dynamics of future music beyond scores and other symbolic approaches.

What is a *gesture*? This is a concept which everyone knows, but no one is able to define. In that way, it is similar to its counterpart, *time*.

The concept of gesture is very important in art, philosophy, and communication. Formally, a *gesture* is a system of continuous curves connecting points in space and time. A curve that connects a gesture to another one is a hypergesture. We explore the creativity of new music which uses the gesture as a starting point for composition.

1.4.3.6 Chapter 15: Future Theories (Counterpoint etc.)

The mathematical approach to music theory is specifically enabling new theories, theories of the future, of perspectives designing where to move.

1.4.4 Part IV: Cultural Extensions

This part discusses cultural extensions of music under the influence of new social and industrial constraints.

1.4.4.1 Chapter 16: A Critique of the Western Concept of Music

This chapter presents a radical critique of the dominant Western concept of music. We question the role of score-driven reproduction of music and its power of abstraction from human reality in the making of organized sounds.

1.4.4.2 Chapter 17: Improvisation and the Synthesis Project on the Presto Software

After a critical review of improvisation, especially in jazz, this chapter deals with the confrontation of mathematical music theory, as implemented in the composition software *presto*®, with the improvisational flight of Mazzola's piano in the piano concerto *Synthesis*.

1.4.4.3 Chapter 18: Art Beyond Music Alone

Music is an art that must be viewed in deep connection to other art forms, such as dance, poetry, or theater. This chapter investigates how art-making methodologies from different disciplines can contribute to each other and suggests that these approaches can be used as modes of academic inquiry.

1.4.4.4 Chapter 19: Human and Machine Music

Music is a genuinely human expression, but it also carries some machine characteristics when misunderstood. They become virulent if reproduction of musical compositions is taken over by humans who function as machines, as perfect puppets. This topic includes a strong critique of so-called Artificial Intelligence, this concept being abused to confuse musical creativity and semiotically flat machine learning algorithms.

1.4.4.5 Chapter 20: Music in Diversifying Cultures

We describe the different ways of understanding traditional music in Ghana for Africa, China and Japan for East Asia, and Indonesia for South Asia. This view is important for a future inclusive treatment of social and cultural musical environments.

1.4.4.6 Chapter 21: Cultural Theories of Gestures

In this chapter, we examine the role of gestures in musical performance. The focus of this examination is on the physical, symbolic, and psychological movements of musicians in different traditions—American Hip Hop, Free Jazz, Japanese Noh Theater, and Chinese Folk. We locate these gestures in the bodies, scores, and sounds of and made by performers. It is also necessary to honor and acknowledge the cultural specificity of each musical tradition by recognizing the importance of different geographies, power dynamics, and histories. We describe the role of gestures as cultural expressivity, complementing the mathematical theory of gestures.

1.4.5 Part V: Creative Strategies

This part deals with perspectives of creativity for a musical future.

1.4.5.1 Chapter 22: Recapitulation of Creativity Theory

This chapter recapitulates our theory of musical creativity to initiate a discourse on how to proceed in musical creativity for the future.

1.4.5.2 Chapter 23: The Specifically Musical Walls Against Creativity

The theory of creativity relies on the concept of walls against creativity. We discuss a number of important walls in the musical realm.

1.4.5.3 Chapter 24: Examples of Creative Extensions in Music

We discuss creative extensions such as new counterpoint, rock jazz, and free jazz, which go beyond the original intention of a genre.

1.4.5.4 Chapter 25: Performance and Composition

We discuss creative extensions by performance of composition and composition of performance for a future enrichment of creative options.

1.4.5.5 Chapter 26: Are Aesthetics and Business Antagonists?

We analyze the problematic but very powerful alliance of artistic expressivity with its commercial framework. For a lot of artists, making money from their art can feel repulsive because (1) selling beauty seems like a contradiction, (2) it is unclear how monetary value can be assigned to an artistic work, and (3) the high art vs. low art, divine versus human paradigm. However, making money from composing is necessary. In this chapter, we will discuss what needs to change to make this more acceptable and look at examples of it working in the past. As artists, we need to remind ourselves that both making and selling are a part of a balanced equation of being a successful artist.

1.4.6 Part VI: COMMUTE

This part presents a short sketch of how a Computational Musical Theory of Everything might look like.

1.4.6.1 Chapter 27: CoMMuTE: Towards a Computational Musical Theory of Everything

This final chapter draws future perspectives of music as a comprising cultural achievement of humans. We discuss the role of music for mathematics and physics from Pythagoras to String Theory, its global human presence, transcending specific fields of knowledge in its synthetical force that unifies distant fields of knowledge and action in the concrete and abstract realms.

Chapter 2
Ontology and Oniontology

This chapter is about ontology of music, including three dimensions: realities, semiotics, and communication. It also includes the extension of ontology to the fourth dimension of embodiment. We call this extension "oniontology" for reasons that will become evident soon.

2.1 Ontology: Where, Why, and How

Ontology is the science of being. We are therefore discussing the ways of being that are shared by music. As shown in Fig. 2.1, we view musical being as spanned by three 'dimensions', i.e., fundamental ways of being. The first one is the dimension of realities. Music has a threefold articulated reality: physics, psychology, and mentality. Mentality means that music has a symbolic reality, which it shares with mathematics. This answers the question of "where" music exists.

The second dimension, semiotics, specifies that musical being is also one of meaningful expression. Music is also an expressive entity. This answers the question of "why" music is so important: it creates meaningful expressions, the signs that point to contents.

The third dimension, communication, stresses the fact that music exists also as a shared being between a sender (usually the composer or musician), the message (typically the composition), and the receiver (the audience). Musical communication answers the question of "how" music exists.

G. Mazzola et al., *The Future of Music*,
https://doi.org/10.1007/978-3-030-39709-8_2

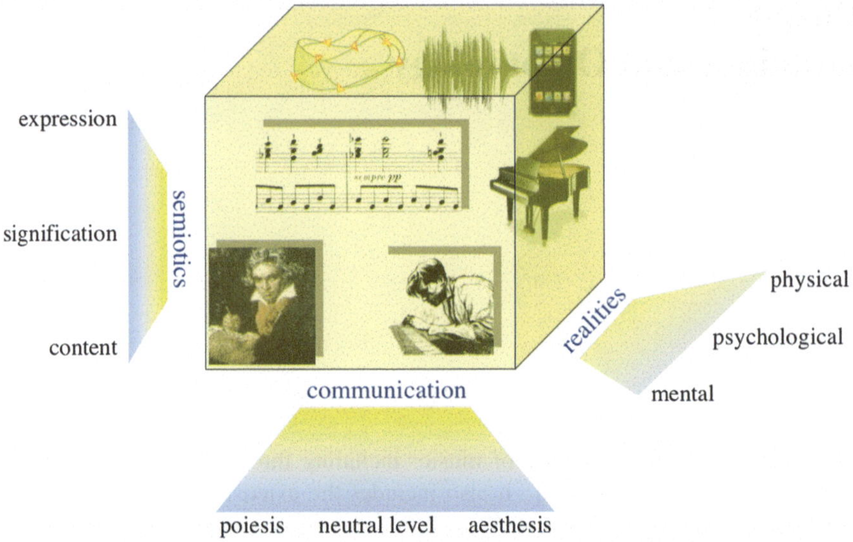

Fig. 2.1 The three-dimensional cube of musical ontology. © Guerino Mazzola

2.2 Oniontology: Facts, Processes, and Gestures

Beyond the three dimensions of ontology, we have to be aware that music is not only a being that is built from facts and finished results. Music is strongly also processual, creative, and living in the very making of sounds. Musical performance is a typical essence of music that lives, especially in the realm of improvisation, while being created. The fourth dimension, *embodiment*, deals with this aspect; it answers the question "how *to come into* being?" It is articulated in three values: facts, processes, and gestures. This fourth dimension of embodiment gives the cube of the three ontological dimensions a threefold aspect: ontology of facts, of processes, and of gestures. This four-dimensional display can be visualized as a threefold imbrication of the ontological cube, and this, as shown in Fig. 2.2, turns out to be a threefold layering, similar to an onion. This is the reason why we coined this structure "oniontology"—it sounds funny, but it is an adequate terminology.

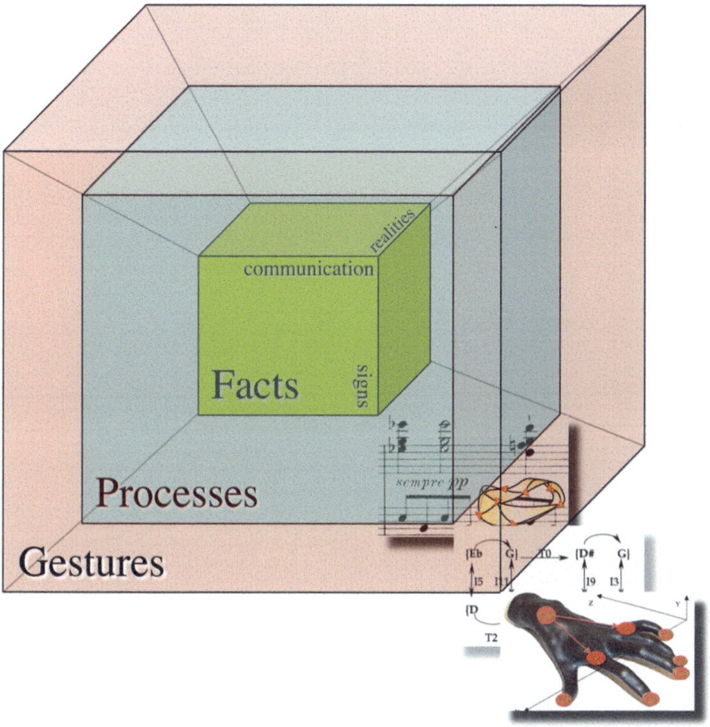

Fig. 2.2 The hypercube of musical oniontology. © Guerino Mazzola

2.3 A Short Characterization

The four dimensions of musical oniontology can be put together to present a short characterization of music:

Music embodies
meaningful communication
and
mediates physically
between its
emotional and symbolic layers.

Chapter 3
The Basic Functions of Music

Music is ubiquitous and it's pull irresistible. Uniquely, the need of music is unlike the need of food; we will not die without it, but we keep being driven to listen to music, or at least, to be exposed to it. There are many ways to access music. For example; attending a performance, going to a club or a restaurant, buying physical recordings (such as CDs, cassettes, and vinyls), accessing a music streaming services, and playing a musical instrument. One might need to listen to music while commuting or driving, to use music as an accompaniment while studying, to watch a performance of already-known or newly-recognized musicians, to get into a certain community where a specific genre of music becomes an identity, or to listen to it in one's leisure time. With several numbers of necessity to access music, we pose a question here: what is the basic functions of music to the human being?

3.1 Surveys of Basic Functions of Music

Scholars have proposed potential functions of music over decades through theoretical point of views and empirical studies. A comprehensive list of theoretical approaches and empirical investigations was surveyed by Thomas Schäfer, Peter Sedlmeier, Christine Städtler, and David Huron [143]. They scrutinized discussions, speculations, and investigations regarding the functions of music listening, and provided a summary as follows.

G. Mazzola et al., *The Future of Music*,
https://doi.org/10.1007/978-3-030-39709-8_3

3.1.1 Theoretical Approaches

In theoretical approaches, functions of music were theorized and deduced from prior theoretical frameworks, but none was the result of empirical testing and exploratory data-collection. These approaches consist of evolutionary and non-evolutionary approaches.

1. **Evolutionary Approaches** These approaches focus on single function of music, comprising as follows:
    - Music projects a social signal of psychological fitness as anyone who can afford the biological luxury of making music exudes vitality (see [124])
    - Music serves a means of social and emotional communication, such as coordinating group activities and reinforcing social bonds, through communicating shades of emotional meaning by the melodic character of emitted sounds as found in work and war songs, lullabies, and national anthems (see [136])
    - Music is intended to maintain infant-mother attachment through humming and singing (see [31, 32])
2. **Non-evolutionary Approaches** These approaches focuses on the needs and concerns of the listeners and attempts to explain how people select and use media in everyday lives to serve these needs, comprising as follow:
    - Music provides ways for entertainment, identity formation, sensation seeking, or culture identification (see [10])
    - Music can be used to activate associations, memories, experiences, moods, and emotions (see [18])

3.1.2 Empirical Approaches

In empirical approaches, functions of music were analyzed through gathering respondent's data. Two goals are intended from the empirical approaches: (1) to observe or identify one or more ways in which music is used in daily life, and (2) to discover potential fundamental dimensions implied by the multiple functions of music. In the latter goal, large numbers of respondent's data are reduced into a few basic dimensions. This will be further discussed within Sect. 3.2. The following is some lists of dimensions of functions of music:

- emotional use, rational use, and background use (see [22])
- ten dimensions according to Boer: emotion, friends, family, venting, background, dancing, focus, values, politics, and culture (see [13])
- eight dimensions according to Lonsdale and North: identity, positive and negative mood management, reminiscing, diversion, arousal, surveillance, and social interaction (see [103])

- six dimensions according to Hays: linking, life events, sharing and connecting, wellbeing, therapeutic benefits, escapism, and spirituality (see [45])

3.1.3 The Comprehensive Empirical Investigations

Schäfer, Sedlmeier, Städtler, and Huron argued that both theoretical and available empirical approaches have some deficiencies. One the one hand, theoretical approaches only focus on single-account function of music, most often because this facet was observed that did not focus on music but focused on other psychological phenomena (e.g. the use of music in managing psychological arousal, the use of music for possible therapeutic functions in clinical settings, etc.). Also, many publications suggest functions of music without providing a clear connection to any theory; some collections remains obscure how the author(s) came up with the functions proposed.

On the other hand, the available empirical investigations, because of using an open approach (i.e. trying to capture the variety of musical functions through surveys and questionnaire studies) or using pre established collections of functions (resulting from specific theoretical approaches or literature research), have yielded quite heterogeneous collections of possible musical functions. Moreover, the overall picture of distilled dimensions of musical functions remains obscure, although at some points they have come to some agreements.

Therefore, Schäfer, Sedlmeier, Städtler, and Huron led an empirical investigations of musical functions, aiming to pursue their analysis without biasing the materials to any particular theory and to use a broad range of all potential musical functions all at once. They identified more than 500 items concerned with the musical use of function, assembled an aggregate list of all of the questions and statements that were theoretically derived or used in previous empirical studies, posing the questions "I listen to music because...", then eliminated and combined redundant items to settle on (eventually) 129 distinct items. After that they categorized the items into three fundamental dimensions. These dimensions are:

1. **Self-awareness** that includes statements about self-related thoughts, emotions and sentiments, absorption, escapism, coping, solace, and meaning. This dimension expresses private relationship with music listening; music assists people to think about who they are, who they would like to be, and how they manage their own path.
2. **Social-relatedness** that includes statements about social bonding and affiliation. This dimension articulates the use of music for feeling close to friends, for expressing one's identity and values to others, and to gather information related to one's social environment.
3. **Arousal and mood regulation** that includes statements about the use of music as background pastime and deviation and as a means to get into a positive mood and regulate one's psychological arousal.

3.2 Music and the Hippocampal Gate Function

It is common knowledge that music and emotions are intimately related. The results of the depth EEG conducted by Heinz-Gregor Wieser, Guerino Mazzola, and their collaborators from 1984 to 1988 suggest a mechanism that could explain this relation on the neurophysiological and cognitive level. Through the test, it can be seen that the emotional brain in its hippocampal structures has a pronounced response to elementary structures of harmony and counterpoints: the intervals in their consonance-dissonance dichotomy in the western classical harmony. Now, the classical thesis of Papez and MacLean [104] states that the limbic system, a prominent part of the archicortex, is responsible for emotional human behavior, this is why it is also called the emotional brain. So the hippocampal sensitivity to consonances vs. dissonances could relate to the emotional function of music, i.e., of musical intervals in our case. The question is, how musical signs which are by no means emotions by themselves can evoke and signify emotions in humans, and why this is done in such a way that the same music may evoke a great variety of such reactions and significations. Evidently these outputs are the result of a determined sample of music plus an individual human ingredient.

The point is that the hippocampal formation has been recognized as a key structure for memory [151]. The neuroscientist Jonathan Winson has proposed a more specific theory of the hippocampal memory function [158], in that he argues that the hippocampus performs a gate function to the subconscious, i.e. to memory contents of emotional character. This means that the hippocampus is a structure that plays the role of a gateway to hidden memory contents. It is well known that humans do not have a free or controlled access to their memory contents, in particular not on the level of long-term and emotional memory, concerning early childhood, for example.

This suggests that special mechanisms must be activated in order to open the hippocampal gate to unveil locked memory contents. It is straightforward from our neurophysiological findings and the gate function of the hippocampus that its musical stimulation could yield such a "key" to open the gate to hidden memory contents. If this were the case, two specifics of the relation of music and emotion would be explained at once:

1. The emotional contents are not generated by music, they are merely retrieved and evoked from a memory database, whence the individual emotional response to one and the same music would receive a logical explanation.
2. It is highly unlikely that the musical stimulation of the hippocampus is independent of the human individual who undergoes this process. In other words, if the music is a key, each individual is likely to have his/her individual key to unlock the "subconscious". This would explain why there are so many different musical tastes—beyond musical education and culture. This would also explain why it is often a specific tune or musical mood that is the personal preference: If this tune played a role in the encoding of a specific emotional memory content, the same key-tune could play a role in the decoding process.

Summarizing, we may conclude that consonant and dissonant intervals and associated harmonic or contrapuntal structures evoke a hippocampus state process which activates a gateway to mainly subconscious memory contents. In other words, Winson's gate hypothesis of the hippocampal formation must also be stated in the sense of the existence of a musicogenic key to the gate.

This thesis does not mean that music produces emotions. We state that it only retrieves and reactivates them from a memory database; it acts on the brain like a drug and produces psychic effects. In this metaphor, the 'chemical formula' of the music drug corresponds to the involved musical structure.

Chapter 4
Historicity in Music

4.1 The System of Music and Its History

In all fields of knowledge, there is a historical and a systematic perspective. The former describes the evolution of the field in time, whereas the latter describes the field's determinants which could persist along the temporal evolution and define its fundamental, systematic objectives. However, these two aspects mostly interact: the systematic part may change along time, while the historical development may be driven by strong systematic vectors.

In physics, for example, the systematic shape which was defined by the Aristotelian writings until the beginning seventeenth century, changed dramatically with Galileo's introduction of the experimental and mathematical paradigm. And nowadays, the theoretical physics more and more dominates the experimental setup. The search for a ToE (Theory of Everything, see Chap. 27) is a purely theoretical axiom, no experimental evidence thereof has been given. But the history of physics is still a purely historiographic fact: physics is mainly the evolution of a big system, not the anecdotal sequence of isolated events. Its history is the trace of abstract ideas of conceptual nature, Newtonian time and gravitation, Einstein's time relativity and curved space-time of gravitation, etc.

Philosophy, to mention a very different case, is not the history of big systems, but can be identified with its history of continuously deepened question complexes. Philosophical systems are a sidetrack of this field, where questioning everything testifies the unfolding of a radical critique of one's own reflection. Philosopher Georg Wilhelm Friedrich Hegel argued that this world is the history of an ever waking up "Weltgeist" (world-spirit), Friedrich Nietzsche taught us to "learn to swim", and postmodern philosophy of Jacques Derrida argued for what is termed "deconstruction".

G. Mazzola et al., *The Future of Music*,
https://doi.org/10.1007/978-3-030-39709-8_4

So where can we locate music in its historic and systematic coordinates? Is it like physics the evolution of one big system of abstract ideas or is it identical with its historic sequence of never-ending variations of the same like philosophy? And what is its role with respect to other fields of knowledge?

To begin with, music is the characteristic art of time, more even than dance since its embodiment is not strictly dependent upon physical bodies. Music not only describes an objectively given time dimension like physics, it also creates time: making (composing/improvising/performing) music establishes a time structure and therefore—following its very logic—time must play a different role also in the music's historical unfolding. Let us have a look at a number of consequences of this musical principle.

4.2 Utopia

A first consequence of the creative relation to time in music is that it relativizes its own history: musical creators often imagine an utopian world: the world of a far future. Composer Karlheinz Stockhausen (1928–2007) viewed himself as being sent from a far star: "I was educated at Sirius and want to return to there, although I am still living in Kü near Cologne." And he added his understanding of the universe: "Why should these large celestial bodies exist if they do not stand for something? I cannot imagine that there is anything senseless in the universe. There is much we do not understand." Stockhausen's creativity was always utopian, the present time was for him a small spot within a big plan. And jazz genius John Coltrane was transcending the present time through Buddhist and similar universalist *weltanschauungen* (world views), also interpreting compositions such as "Out of this World", or transcending present musical limits in collective improvisations such as "Om". Jazz composer Sun Ra (1914–1993) insisted on his origin from Saturn, he viewed his earthly presence as a small event within an eternal thread of cosmic art. The titles of his art are typical for this attitude, e.g., "The Futuristic Sounds of Sun Ra" is an album recorded October 10, 1961 for the Savoy label.

It is interesting to notice that a famous utopist, the French sociologist and economist Charles Fourier (1772–1837, not to be confused with Joseph Fourier, the physicist) wrote an utopian view of humanity [34] where humans are freed from normal work and can enjoy food and make love and are also emancipated from the inequalities of gender and race.[1] The remarkable characteristic of this utopian perspective is that Fourier classifies his societal structures according to musical principles. On page 341 of the second volume, he describes cosmic principles in these words[2]: "Mercure, 5^e satellite de la terre, lune favorite ou rectrice e l'octave

[1]Fourier seems to be the creator of the idea of women's emancipation, more than 180 years ago!

[2]Mercury, fifth Earth satellite, favorite moon or rector and major octave.

majeure (...)" All of Fourier's principles are musically driven. His harmony is a musical utopia.

The musical utopia was also stressed by Richard Wagner's idea of a music of the future in his essay "The Artwork of the Future", published 1849. In this work, he proposes an artwork of the future, which would comprise music, dance, and poetry, and would converge in what he called "Gesamtkunstwerk" (total work of art), a concept which anticipated the idea multimedia art.

The futuristic ideas in music also became virulent in the first decades of the twentieth century and was coined "Neue Musik" (new music) by the German conductor and music critic Paul Bekker in 1919. New music includes the search for new sounds, new forms or combinations of old styles, either through the continuation of existing traditions, or through the break with tradition. Journal titles such as "Perspectives of New Music" are symptomatic for this spirit.

These attitudes prove that the future has often been a standard topic in music's history. The musical characteristic is also a historical feature. But what about the past music? How does the past of musical creativity relate to the present and future? Looking at the recent history of musical developments, one observes that the past has also been revitalized. For example, orchestras with original eighteenth century instruments,[3] such as propagated by Nikolaus Harnoncourt, were introduced, and the oxymoron of "New Tonality", "New Simplicity" were introduced by composers such as Arvo Pärt or Krzysztof Penderecki, who, after serial experiments returned to tonal composition. And the famous "Switched-On Bach" studio album by the American composer Wendy Carlos, released in October 1968 by Columbia Records, testifies that new perspectives (this time electronic sounds) of old music are always a creative contribution to deepen what was thought to be completely understood.

But going back is also an essential tool in composition! The recapitulation in the sonata form means looking back and expressing the changes which occurred in the development when playing the exposition a second time: Time is recycled in this very classical method. In other words, the historical dimension in music is different from the case of physics or philosophy. Music has, by its very internal logic, a creative relationship to past and future.

Does this mean that there is no systematic layer in the history of music, i.e., that music is always recycling the past and anticipating the future like philosophy, and there is no deeper systematic level of the "science and art of music"? No, this is too simple. Music has always had a strong theory, from Pythagorean intervals to contrapuntal theories of Johann Joseph Fux or Hugo Riemann's harmony, but those theories developed more in the spirit of mathematics than in the physical paradigm. Let us make this clear. Mathematics also develops, and sometimes with a giant steps. But mathematical theorems are never replaced by new theorems because they turn out to be wrong. They are only replaced by theorems which cover deeper structural insights.

---

[3]Observe that playing classical early nineteenth century composers on twentieth century instruments is also a synthesis of different musical times.

Music evolves more in this style: deepening insights, not replacing results because they are wrong. This is a clear difference to astronomy, where some experimental results may enforce a radical revision of an outdated theory. For example, the principle of circular orbits of planets had to be replaced by elliptic orbits in Kepler's theory. But musical historicity is also different from mathematics, where the shaping of time is not a relevant category, neither historically nor within the mathematical processuality.

4.3 Musical Anticipation

The history of music has a remarkable interaction with the development of other sciences. It is interesting, for example, to list all the points where music and its theory were in advance of physics:

1. Pythagorean world formula of tetractys was a musical paradigm that could be reified only much later, in the theoretical physics of the twentieth century.
2. Gesture theory in music was developed much earlier than physical string theory. Gestures are essential theoretical concepts in performance theory as described by Graeser, Adorno, Sessions, and Lewin.
3. Kepler's laws describing the movement of the planets were musically motivated, he was in search of musical interval fractions (3/2 for the fifth, etc.) to understand astronomy.
4. In music theory, the three basic harmonic intervals, octave, third, and fifth, were conceived as independent harmonic directions, an interpretation which could later be confirmed mathematically when mathematicians had introduced the concept of linear independence in linear algebra.
5. Grothendieck's mathematical topos theory was applied by Mazzola to music theory before the physical application of topos theory took place, essentially by the work of mathematical physicist John Carlos Baez, a cousin of singer Joan Baez.
6. Symmetries as fundamental tools for music theory were applied by music theorist and mathematician Wolfgang Graeser 4 years earlier than in physics by Emmy Noether.
7. And a sad anticipation: Music business was destroyed earlier than the open source catastrophe in the sciences.
8. Contrapuntal experiments in musical composition were executed since the early polyphony around 900 AD, before Galilei's approaches took place in physics in the seventeenth century.

Chapter 5
Only One Restriction: Quality

5.1 Intellectual Properties in Music

The concept behind our first step in objectively determining the quality in music is created from our ability to discern how much intellectual effort is needed to recreate the same work. Through analysis, we can determine that the composer uses different theoretical concepts from the traditional Western harmony canon to create their work.

It is imperative to acknowledge our first caveat in the realization of intellectual property in music. There will oftentimes be situations in which the audience member cannot truly determine the extent of intellectual property in the quality. In Fig. 5.1 a collection of notes that, to a trained musician, would assume were selected by an excitable child at random. Only through an in-depth analysis, one realizes a great deal of time was afforded to the selection of these notes by composer Pierre Boulez. For a more in depth analysis of Pierre Boulez, see Chap. 11.

Though it is relevant to recognize the structural complexity in creation, it is more important to acknowledge the application of theoretical knowledge. We, as listeners, need to realize there is a language-like quality to music. In certain cases, our age may determine how accurately we are able to perceive the intellectual properties presented in music.

Neuroscience has shown that there is a critical window in the development of humans, in which as time progresses through maturation, a definitive correlation is apparent between age and the ability to learn language. Scientists do not dispute the correlation, but are still attempting to discern the reasoning as to why it works in this fashion.

This relationship is relevant because spoken language, to an extent, is a collection of sounds. As we mature, our ability to recognize sound begins to deteriorate unless we are constantly exposed to sounds that are not usually around us. The speakers of tonal languages, such as Chinese, are more likely to possess perfect pitch which stems from semantic differences by varying the intonation given to words or

G. Mazzola et al., *The Future of Music*,
https://doi.org/10.1007/978-3-030-39709-8_5

Fig. 5.1 A collection of notes. © Christopher Rochester

syllables of a similar sound. The reason lies in the concept of the degree to which they have been exposed to the intellectual properties in speech which translates to an understanding of tonal nuances in music.

As research continues into child development in neuroscience, we will eventually have a greater ability to perceive the intellectual properties of music.

5.2 Communication of a Musical Message

We cannot focus solely on the application of theory as a criterion in determining the worth of music. The communication of the musical message, or the relationship between the poesis and aesthesis is instrumental in the evaluation of quality in music. Oftentimes, a perceived lack of the application of music theoretical knowledge can be eclipsed by the success brought forth in the communication of a thought-provoking idea.

African American work songs used by the Underground Railroad during slavery are a powerful example of vital musical messages. Many of these compositions contained coded directions or warnings for escaping slaves. "Wade in the Water" let people know to get in the water to avoid being seen and "Sweet Chariot" alerted individuals to prepare to escape. Musical composition was considered in the creation but the primary importance of these work songs is the message conveyed to listeners and the history that accompanies them.

Coded messages have been a part of African American music since work songs and emerged in jazz, blues, soul, funk, and Hip Hop. Most famously, Chuck D of Public Enemy has said, "Rap is the CNN of young Black America" which refers to how information is spread through Hip Hop music and calls attention to racist framings of news reports about African Americans. Although Hip Hop music in particular is derided as lacking complex musicality, many groups like Digable Planets, Ice-T, and RUN-DMC crossed genres to create rap rock, rap metal, and jazz rap.

5.3 Medium

Instrumentation factors in heavily in the execution of a work... A work song performed by a group of MIDI instruments would result in a completely different reaction from the viewpoint of the aesthesis. The use of this example is not to denounce the relevance of MIDI instruments... Rather, it is an acknowledgement of the idea that there is an important relationship between the artist's medium and audience.

We must acknowledge that our criteria of quality cannot devalue fundamental aesthetics and definitions in non-classical forms of music. Hip Hop DJs innovated the scratch as a foundational technique in the art of turntablism but are considered arrangers and not composers through the lens of classical music. For example, Q-Bert suggests the existence of at least twenty-five different scratches such as the Bubble Scratch, Crab Scratch, and the Hydroplane ("A Taxonomy of Scratching" [91]). How do we translate the compositions of scratches which have been validated by musicologists and musicians as legitimate in academic settings ("A Scratch Composition" [92])? Sampling is another technique from this musical form that is derided because of the use of pre-recorded tracks. Can we consider this prejudice as lock-stepped in an acoustic mind-set of Tradition? It is well-known that even Mozart sampled popular songs of the time and incorporated these harmonies into his most famous compositions. What is the difference between concepts and approaches of non-traditional techniques in varying musical forms? Ideas about instrumentation must, for the sake of continued relevance beyond the ivory tower, regard ways of composing that differ from the entrenched classical tradition as important and valuable to the field of music study. While this distinction is important, we do recognize that Hip Hop turntablism is a different medium than acoustic instrumentation and requires a different skill set.

5.4 Cultural Factor/Relevance

Our quality is primarily determined by the combination of these factors. We have attempted to separate the cultural relevance as a means of determining quality in this situation. The contributions of people of color have been largely disregarded in popular music pedagogy. Consider the concept that there is music that does not necessarily perform well within our set criterion, but this same music has the ability to enhance its own quality when we acknowledge its cultural importance.

Beethoven's influence on the classical genre has spanned generations of musicians and his music is regarded as a pinnacle of excellence in the field. Comparatively, Tupac Shakur, a highly influential Hip Hop artist, whose enormous global impact on society can only be measured in orders of magnitudes in issues of social justice with direct activism and resistant musical content [152]. While Hip Hop music continuously garners contempt from academics as musically simplistic, the reception of Tupac's music must be considered as it is: historically, culturally, and socially situated and relevant. Perhaps much more relevant to the present-day than Beethoven, Tupac, by virtue of his profession as rapper, is not held in esteem by those who practice music derived from the traditional Western canon. The reason for this inequality is based on the composition of the music and not the social influence of the artist.

Beethoven's fifth Symphony is one of the most well-known compositions by the artist and has become embedded in the teaching of classical musical theory. His brilliance cannot be denied but does elide a preference for European canon over all others in considerations of musical "quality." Referring to the earlier criteria versus 2Pac California Love.

5.5 Quality

Quality is a factor in music in which all composers should aim to improve. It is important to acknowledge that the key difference between audience members will ultimately be their apophantic relationship with sound. Many years ago, human survival was completely contingent on our ancestor's abilities to discern patterns in their surrounding environments. This pattern recognition has been essential in our evolution and is evident in our evolution.

Pareidolia is a type of apophenia that refers to the concept of seeing patterns in random data. The most common example being the Rorschach test, a psychological test in which a subject's perception of an object is recorded and analyzed. This same conceptual idea of seeing patterns in art should correspond with the idea of hearing patterns in music. That is to say, it is part of human nature to perceive patterns in their everyday occurrences.

In Fig. 5.2 you as the audience, have been given the freedom to create your own narrative in constructing the identity of the inkblot. In music we are given the same

Fig. 5.2 Inkblot from Rorschach test

opportunity, which ultimately is the reason people have different relationships with music.

Music, itself is ambiguous in nature. We can romanticise the relationship between the music and the aesthesis as one in which the foundation is uncertainty. The notion of quality is not an idea in which we should use to criticize concepts of the past, but a tool for future composers to elevate music to a new level of expressivity.

Part II
Technological Tools

Chapter 6
Software Tools and Hardware Options

6.1 Composition and Sound Synthesis: DAWs

DAW is the abbreviation for Digital Audio Workstation. Most DAWs have functionalities such as recording, multi-track editing, mixing, and MIDI programming. The functionalities of DAWs have much improved and extended in the decades with the ability to support third-party plug-ins, audio sample libraries and either software or hardware samplers. Video synchronization is also a trending feature which is specially designed for film scoring. Meanwhile, the score editing feature is preserved for composers who prefer the traditional way of writing.

6.1.1 Recording

This is one of the key features of DAW which allows musicians to record live instruments and voice. To enable recording, a suitable recording environment, competent recording devices and a sound card with Analog-to-Digital Converter (ADC) are necessary for the DAW to capture sound generated in the air. The analog signal is then transformed to digital signal, or in another word, a glut of discrete-time and discrete-amplitude samples of continuous signal, through ADC, and be further delivered to the DAW itself. This transformation process is designed in a way in which enough samples are maintained to convince human ears that it is the same sound originated from the analog signal.

Once the DAW gets the digital signal, it starts processing it and visualizing it on the screen by waveform. This usually happens in real-time and can be monitored by devices like speakers etc.

G. Mazzola et al., *The Future of Music*,
https://doi.org/10.1007/978-3-030-39709-8_6

6.1.2 Project Managing/General Editing

Nowadays, managing a large-scale project and implementing complex ideas in a piece with DAWs is relatively simple. Media meta information such as tempo, time signature, length, can be set and modified whenever necessary during the composition. And it's also possible to change them with functionalities such as tempo/time-signature markers, tempo tracks (utilized curve to shift tempo). There are also many features help users to compose conceptually, visually, logically, or with certain sound aid. One worth noting conceptual project managing feature is the idea of region or pattern combinations. For instance, *FL Studios* [64] has a pattern system to allow users to create a snippet of notes which could be reused as a repetitive element in the piece, see Fig. 6.1.

Figure 6.2 demonstrates the tracks inside an FL Studios pattern. Not all tracks have to have content in a certain pattern.

The same goes with *Reaper*, which allows users to define, move and copy regions which contain multiple tracks of notes. Visually, many DAWs support color system, instruments icons to help users to identify tracks. Some DAWs has built-in visual representation of tracks such as *Logic Pro* [72], and some DAWs support color system customization and instruments icon customization, like Reaper [84], see Fig. 6.3.

Also, it's possible for users to create logical tree structure in terms of track management in which the tracks are organized in a hierarchy; Fig. 6.4 presents two families with subtracks.

For example, users can have violins, violas, cellos, basses under string section, by doing so, it is possible to manage and edit in both section-wise or instrument-

Fig. 6.1 FL Studios: Each block in the figure is a pattern. This figure presents how different patterns are utilized and combined into music

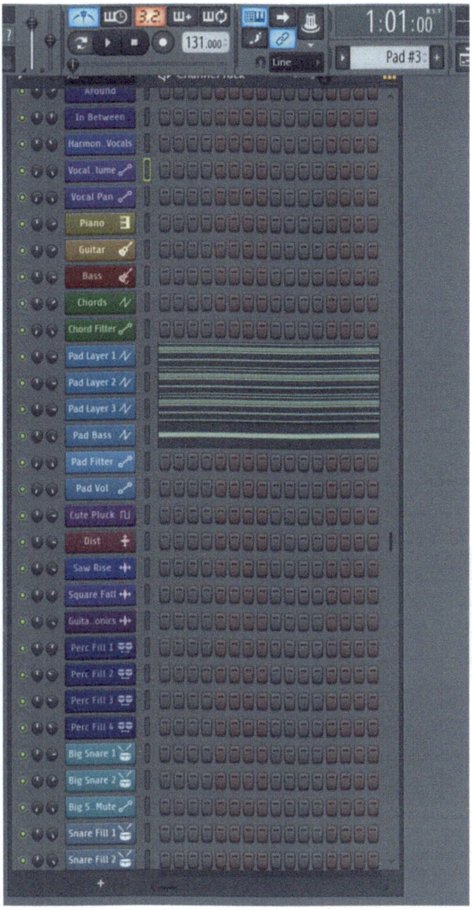

Fig. 6.2 FL Studios: Tracks inside a pattern. Not all tracks have to have content in a certain pattern

wise. With film scoring project, multi-layer sub-project system comes into play which share similar ideas as multi-layer track system mentioned above.

6.1.3 Audio Editing

Manipulating volumes is achievable by functionalities such as cross-fade between audio clips, track volume controls, and volume envelope.[1] It is also possible to time-stretch audio clips. Different DAWs might have different ways to achieve manipulation of volumes, but the general idea is the same. In Reaper, one can use

[1]An envelope is a function of time whose value adjusts a given parameter, such as volume, tempo, or pitch.

Fig. 6.3 Reaper: The personalized color scheme and icons

stretch markers and different key combinations to quick stench and shift a certain moment or small area inside an audio clip. Of course, adjusting the whole audio clip is doable by setting the property of the particular audio clip that one might want to adjust.

Pitch shifting, and correction is also an important audio editing feature for recordings. By either editing the audio property or adding a pitch shifting plug-in to the track, one will be able to change the key within a second. Also DAWs' build-in pitch correction tools or third-party pitch correction plug-in do the same trick.

Fig. 6.4 Reaper: Two families with subtracks

6.1.4 MIDI Programming/Recording

Programming using the MIDI code is supported by DAWs such as *Pro Tools* [82], Logic Pro, Reaper and *Cubase* [58]. MIDI is the abbreviation of Music Instrument Digital Interface and was introduced in 1983 by the NAMM (National American Music Merchants) as industrial music communication code, see [75] and [127]. It usually supports pitch, onset, offset, loudness.[2] In other words, MIDI can be understood as the 'score' for digital devices. Composers can access to piano rolls and input MIDI notes either by hardware keyboards or manual input, see Fig. 6.5.

Reaper supports multiple recording modes such as MIDI over-dub or MIDI replace. The former enables users to record MIDI information and combine different recordings while the latter allows users to replace an unwanted recording. To ensure the preciseness of the playback, many DAWs also support quantization of notes by aligning midi input to a grid interval of one's choice.

To further control MIDI, (Continuous Controllers) are essential tools for realizing the musical articulation. For instance, CC64 is a special MIDI controller for instruments with a sustain pedal. CC7 is usually used for volume control. Users

[2]The term *velocity* is commonly used in MIDI.

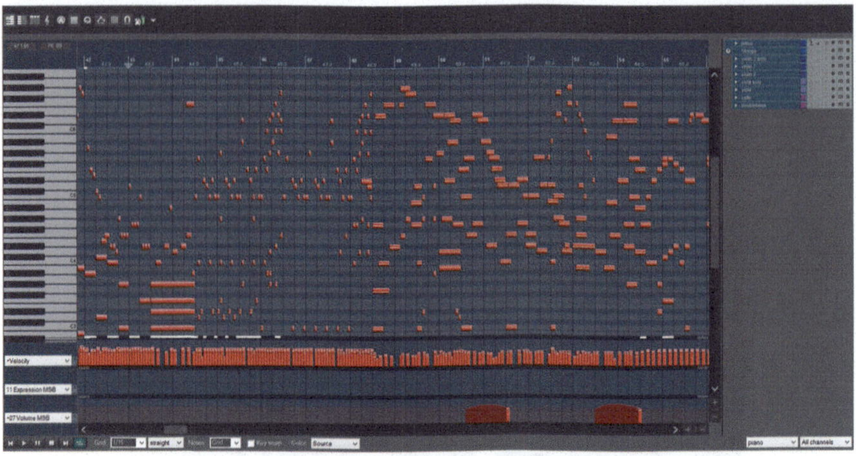

Fig. 6.5 A typical piano roll in DAWs

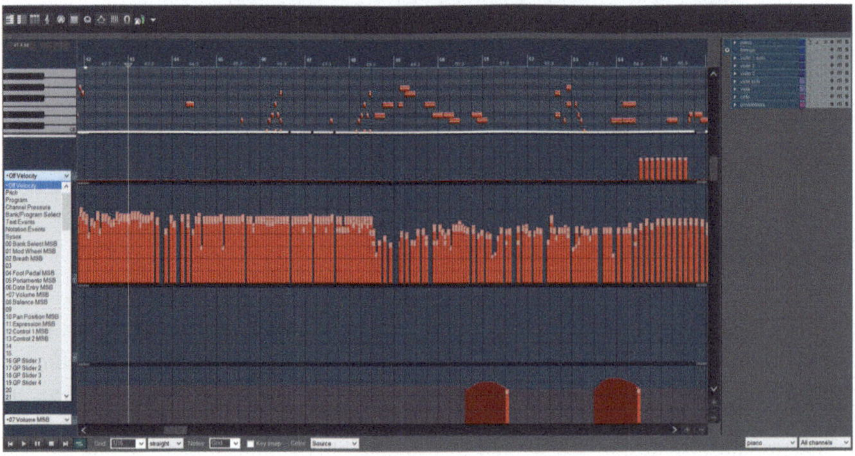

Fig. 6.6 Reaper: A list of continuous controllers (CC)

can customize CC controllers as well, and many third-party sample libraries also have special utilizations with CC to control the samples they provided. Figure 6.6 presents a list of cc controllers.

6.1.5 *Mixing/Mastering*

Software mixers are embedded in most DAWs. They free modern composers from relying on expensive hardware to create an audio outcome of their composition.

There are many aspects of mixers in DAWs, such as FX plug-ins, channel routing, mixer controllers, and automation. FX plug-in is used to apply different mixing effects to a track. For instance, equalizer is a typical type of mixing FX plug-ins which can adjust the frequency content of audio. Many DAWs have built-in FX plug-ins or one can use third party plug-ins to help balancing the frequency content of a track.

A channel routing system is a crucial functionality for both audio tracks and MIDI tracks. With a routing system, it is possible to send sampler processed MIDI information (with the sound of sample) to an audio track by which the MIDI information becomes a virtual audio output. It is a simulation of routing between hardware FX plug-ins and physical audio devices. Also, the routing system enables processing audio tracks by multiple FX plug-ins in different tracks or sharing FX plug-ins between audio sources.

Mixer controllers are general controllers inside a mixer, which include volume control, mute button, FX plug-ins button, and routing control. They are interfaces to mix related functionalities of a track.

DAW automation[3] is a heavily used feature in mixing, which controls the degree and variation of continuous and discrete parameters. Automation can control general properties, such as volume of a track or a single parameter of an FX plug-in.

6.2 Composition and Sound Synthesis: Audio Programming

In addition to DAWs, audio programming can also be used for composition and sound synthesis. Audio programming serves customization for one or more computing tasks such as sound synthesis, algorithmic composition, designing virtual architecture (e.g. virtual routing), creating a GUI (Graphical User Interface), and analysis. Those tasks are performed through inputting a set of instructions (i.e. program) into a computer; a program is written in a formal language which comprises of words using letters from an alphabet and is syntactically well-constructed. Because of this nature, consequently, any errors made in inputting a set of instructions may cause an inability for a computer to execute a certain task.

Despite the drawback, audio programming has some advantages. For example, it enables a computer to perform sound events that might be impossible for an acoustic instrument to do (e.g. the events may contain extremely short durations, extremely large frequency intervals, or extreme timbre range), be an extension of an acoustic instrument in that a computer manipulates sound input of an acoustic instrument (or any physical things that can be used as an instrument), and present a customized application for a specific purpose.

The language used in the recent development of audio programming is written in high-level scripting languages which is more human-readable as opposed to low-

[3]Envelopes are typical automations.

level (machine) language which provides little to no abstraction making it difficult for us to read. The high-level scripting language in audio programming is presented either in graphical (or visual) programming or text-based programming language, and is widely based on the OOP (Object-Oriented Programming) paradigm.

6.2.1 Object-Oriented Programming and Its Advantages

(Object-Oriented Programming) The OOP classifies data bundles into *objects*; an *object* contains *states* and *behaviors*. An *object* stores its internal and private states into *fields* (or *variables*); its behaviors are declared through *methods* with specific access modes. A *class* models *states* and *behaviors*; it presents a blueprint or prototype of an *object*. An interaction done by *methods* towards private *states* is called *data encapsulation*. This nature provides the condition, called *abstraction*, where one is only exposed with a high-level mechanism, i.e., one only has to send *messages* to modify an *object*'s *state*. The messages may be in the form of primitive data types (such as Boolean (true or false), integer, real numbers, characters, or strings) or another *object* [69–71, 79, 154].

To understand the OOP more, consider this real-world example. A *class* piano has *fields* such as the keys, the strings, and the sustain pedal. These states can be accessed through public *methods*, for example, press selected key(s), release selected key(s), pluck selected string(s), and hold a sustain pedal, respectively (see Fig. 6.7). To play the *class* piano, an interaction between *fields* and *methods* is required (*data encapsulation*), and one only has to go through a high-level mechanism by modifying the piano's *state* or, in other words, sending *messages* to them (*abstraction*).

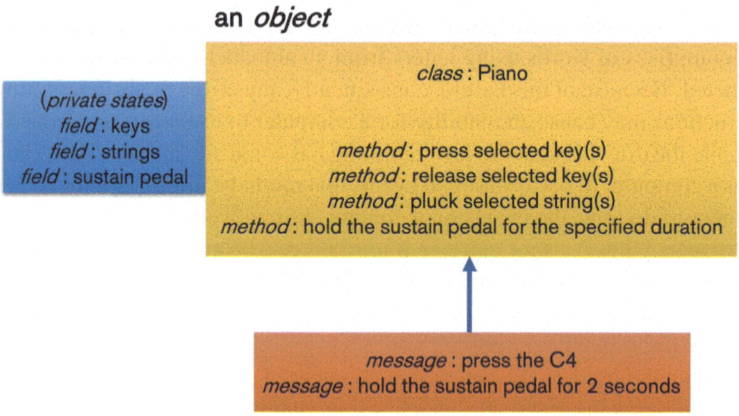

Fig. 6.7 The *class* piano

This paradigm might be what makes the OOP popular; we only have to access the *methods* and 'do not have to take care of' the private *states* and their complicated architecture. Although users still have to understand programming syntax, the nature of the OOP and the fact that myriad *classes* provided by most audio programming platforms enable users to make an advanced prototype considerably quicker become advantages for musicians who do not have a strong programming background. Here is a list of well-known OOP languages (some of them extensions of the classical non-OOP Ansi C language): C++, Java, Objective-C, Swift, Python, Smalltalk. Java is platform-independent, Python works similar to Java on many platforms, C++ works on Microsoft and IBM, while Objective-C and Swift work on Apple computers.

6.2.2 Graphical Programming Language

A graphical programming language uses a data-flow paradigm in that basic functional units connected with patching 'cords' are used as an interface. Those basic building blocks, called *objects*, act differently depending on data types; there are integers, real numbers, Boolean's true/false (may be presented in the form of an "on/off" button), strings, a user interface, and a *class*—in fact, every unit is called an *object* in graphical programming language regardless whether it is a *class* or a primitive data type. This type of language works serially in which data is read from an object to another object.

Max (or widely known as Max/MSP) [73] and *Pure Data* (or Pd) [83] are two popular graphical programming languages used for audio programming, see Fig. 6.8 for a MAX interface. Both Max/MSP and Pd are capable for real-time audio synthesis, enabling us to edit the program while it is instructing a computer to execute certain tasks without interrupting its performance. Max/MSP and Pd are capable to work with MIDI and audio as well as to program 2D and 3D visual elements (Fig. 6.9).

6.2.3 Text-Based Programming Language

A text-based programming language uses text to program audio. This type of language works serially in which data is read line by line. Some audio programming platforms have ready-to-use classes to create from a simple to an advanced realtime synthesis (e.g. *CSound* [57], *SuperCollider* (SC) [88], and *Chuck* [54]), some platforms mainly support live coding that provides simplicity in language (e.g. *Sonic Pi* [87], *TidalCycles* [90], *Extempore* [61], *ixi-lang* [68], and *Fluxus* [65]), and some platforms support libraries of classes for audio coding (e.g. *STK* for C++ [52], *Gibber* for Javascript [66], and *Overtone* for Clojure [81]). These platforms

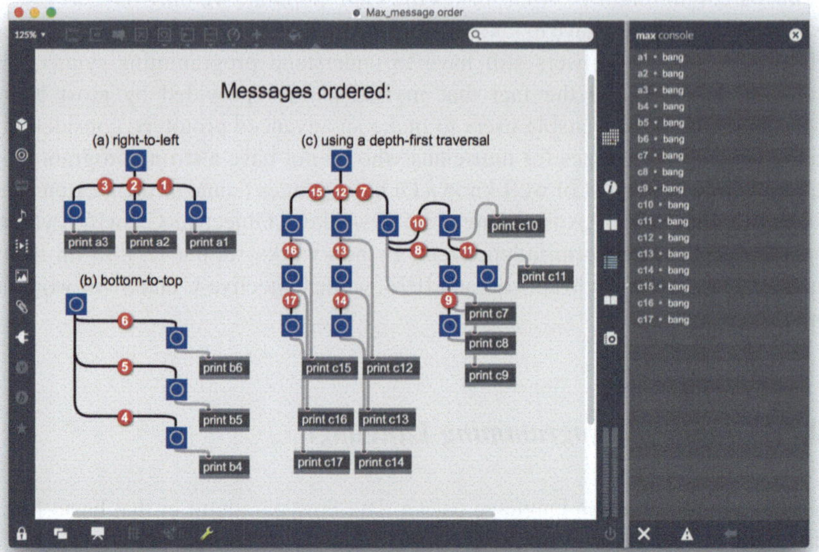

Fig. 6.8 Example MAX interface showing message execution order

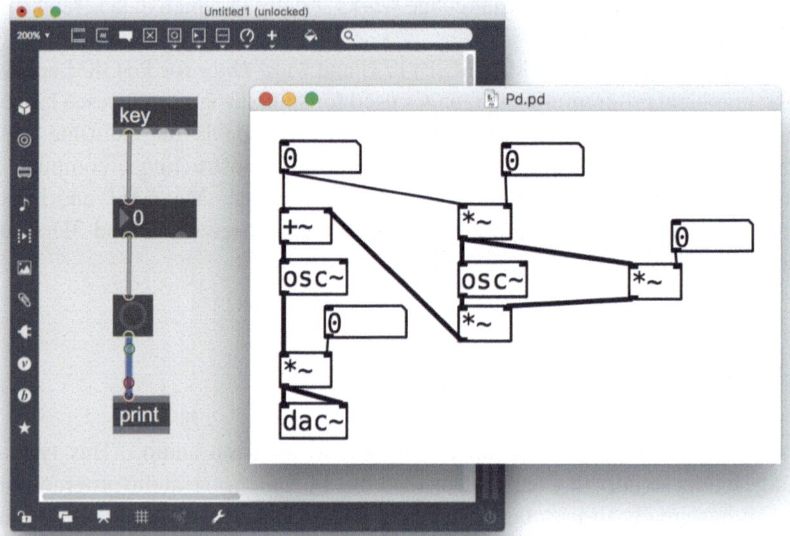

Fig. 6.9 Examples of graphical programming language

are expandable in which one can create and add a special class to serve a certain task.

With the invention of Open Sound Control (OSC)—a protocol for communication among computers, sound synthesizers, and multimedia devices [80], more than one platforms can be connected, either between text-based programming languages or between a graphical and a text-based programming language. For example, SC as a powerful audio synthesis can be treated as an audio engine which is controlled by another platform as an input, e.g., Pd. Even an audio programming language can be connected with other text-based programming languages that is capable for building visuals such as Processing and openFrameworks (Fig. 6.10).

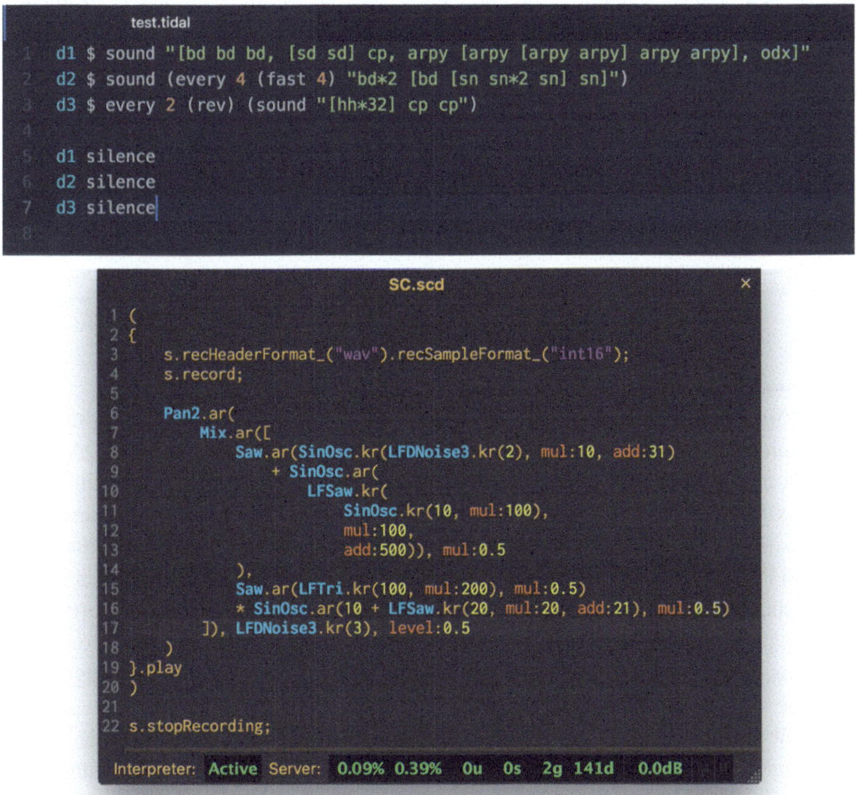

Fig. 6.10 Examples of text-based programming language

6.2.4 Which Language Is Better?

It is sometimes confusing for us when choosing which language is better to begin with. Both languages have advantages and disadvantages.

A graphical programming language provides a visual aid to understand programming flow so that what we see is what we get. This nature also helps musicians who are beginners in programming. However, a visual aid of the language needs more spaces as soon as a patching environment is complex, and this may also mean patching cords can look messy.

On the other hand, a text-based programming language needs less space compared to a graphical one since it relies on high-level abstractions, allowing it to be more expressive. For example, one may only need several lines of code to produce a granular synthesis in a text-based programming language, while one needs big space to create the same synthesis from scratch in a graphical one. However, one should learn a programming syntax to understand how a certain programming language works, and without a visual aid, it is sometimes difficult to understand.

Regardless the advantages and the disadvantages, both language types are capable to do the similar tasks (of course, depending on a certain platform). Choose a platform which you are comfortable with.

6.2.5 Audio Programming and DAW

Audio programming is not meant to substitute DAW as their natures are different. Although both can do similar things in sound synthesis and composition, DAW has a fixation and constancy nature compared to audio programming's customization and flexibility. This, in fact, gives us more choices in creating synthesis and composition. For example, if one wants to do live coding, synthesize sound (e.g. through FM and granular synthesis), chain simultaneous tasks (from analysis, sound synthesis, to flexible controlling), and connecting a programming platform with another platform through OSC, certainly audio programming is a way to do so. Whereas, if one wants to do audio post-production like mastering and manage fixed multiple audio files and MIDI, certainly DAW is a way to do so.

DAW and audio programming also can go hand in hand. For example, DAW can be used to manage audio files that are synthesized through audio programming, like in the case of composing fixed-medium works (fixed or predetermined works bounced in any audio file types that are presented through two to many speakers).

6.3 Analysis and Experiment

6.3.1 Rhythm and Melody Creation via Automata

There is software to generate rhythms and melodies. An example is the work of Stephan and Richard J. Hollos, which develops Ansi C language for generating rhythms [46] and melodies [47]. In their conceptualization, rhythms are n-periodic sequences of onsets in $\mathbb{Z}_n$. They describe different mathematical methods to create special types of rhythms. Melodies are understood as being sequences of pitches at given onsets, so essentially these are subsets in $\mathbb{Z}_n \times \mathbb{Z}_m$, where the first coordinate is a rhythm with period n, while the second coordinate is a pitch class. The methods for melody construction are taken from the theory of formal languages and associated automata. Figure 12.3 shows an automaton from [47] for two octaves of the diatonic scale connected in a circular geometry. The pitches are encoded by encircled numbers (two octaves here, 0–6 and 7–13), the possible transitions are defined by arrows connecting pitches. Observe that this formal approach via automata also defines the possible transitions in first species counterpoint (see Sect. 15.1.3.1). The questionable point is not one of formal methods, but about the choice of automata which are musically effective (Fig. 6.11).

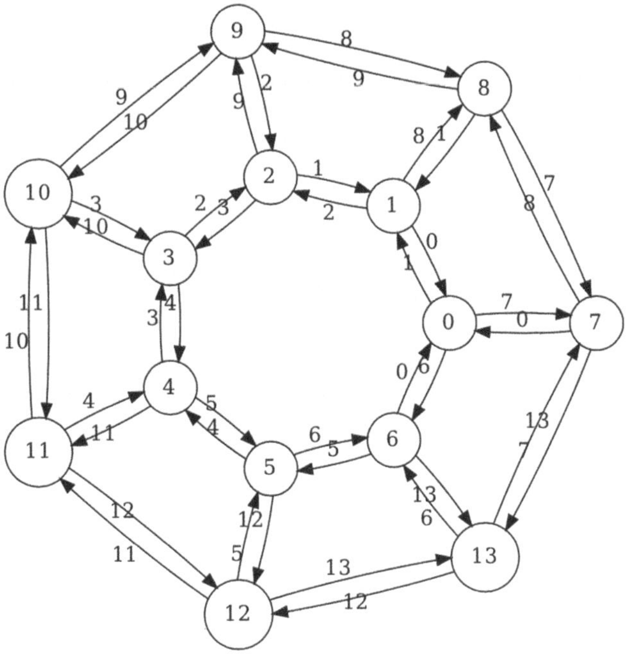

Fig. 6.11 An automaton for two octaves of the diatonic scale connected in a circular geometry

6.3.2 RUBATO® Components for Analysis

The RUBATO® software was implemented since 1992 in Apple's Objective C language, and reimplemented in 2003 in Java to become a platform-independent application. This software has three types of components (called rubettes in this context): composition, analysis, and performance. The original scope of this software construction was to implement computer-generated performance of classical European scores (provided as Standard MIDI files) expressing analytical inputs of rhythmic, melodic, and harmonic nature. All these rubettes would perform their specific analyses, but their output was programmed to be in the form of weights (non-negative real numbers) attributed to all notes of the given score. The reason for this numerical language was that performance cannot directly use abstract analytical symbols, such as Riemann functions Tonic (T), Dominant (D) or Subdominant (S). The performance must rely on numerical information to be applied to musical parameters of tempo, dynamics or articulation.

The rhythmic rubette, called MetroRubette, in its version for the Java RUBATO® is explained in Sect. 6.4.

The melodic analysis rubette, called MeloRubette, would first exhibit the list $MOT(COMP, NUM, DIST)$ of all motives within a defined limit (number NUM of notes and distance $DIST$ between first and last note) on the given score $COMP$. Next, one would define and calculate the weight $w(M)$ of each motif M with regard of its presence within other larger motives and its content of smaller motives. The weight of a note n would then be $w(n) = \sum_{n \in M} w(M)$. This software component has been used to discuss the still controversial question whether the main theme of Bach's *Art of Fugue* would consist of the first eight notes or else include the total of twelve notes. This research revealed that the 12 note version does not show up new motives of high weight, which were not already present in the eight-note motive, see [19] for details. This analysis was performed on the old RUBATO®

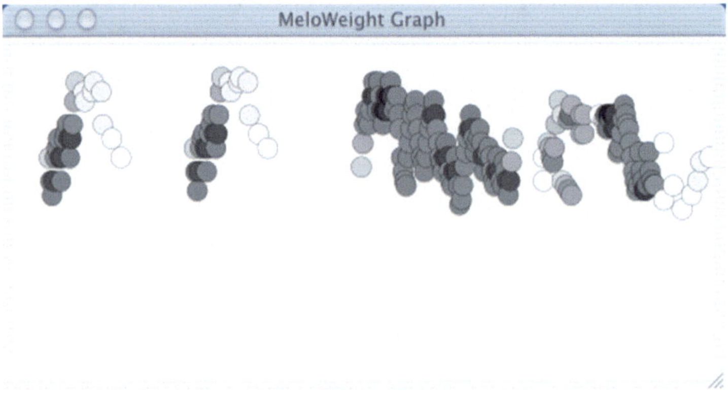

Fig. 6.12 The weight graphics for the celli part 30 in the score denotator deduced from Richard Wagner's *Götterdämmerung*. © Guerino Mazzola

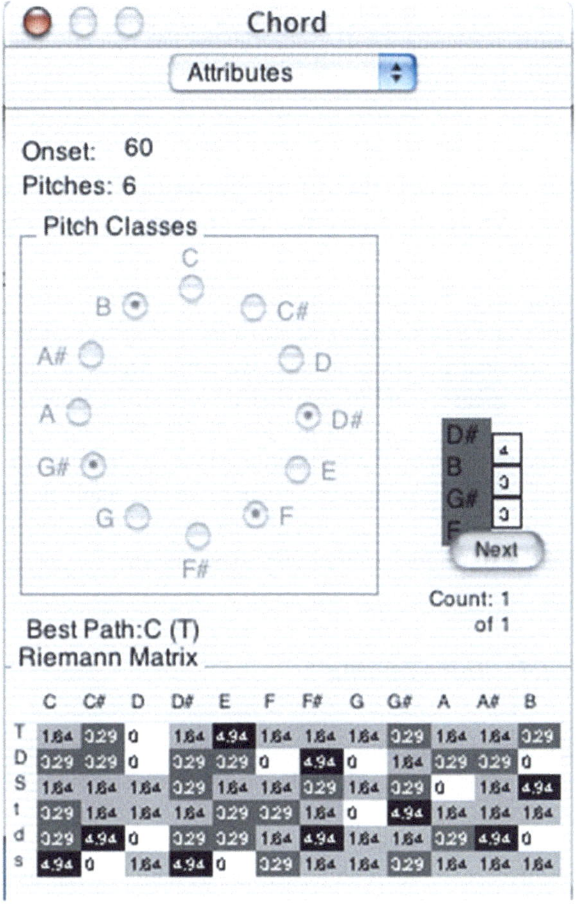

Fig. 6.13 The ChordInspector of the HarmoRubette shows each chord of the chosen chord sequence with its pitch classes, the third chains, and the Riemann matrix. The grey level of the values is proportional to their relative size. © Guerino Mazzola

version, a new MeloRubette is being implemented for the Java RUBATO®, see Fig. 6.12 for a melodic weight graphic, dark color encoding high weight. Details are provided in [118, Ch. 41.2].

The harmonic analysis rubette, called HarmoRubette, used Riemann function theory concepts, Tonic, Dominant, and Subdominant, to create harmonic note weights. The method worked by generating for every chord of the composition a "Riemann matrix". This matrix would give every potential tonality $C, C\sharp, D, \ldots B$ and mode (major, minor), a Riemann function value for each of the three Riemann functions. The important difference of this concept is that the harmonic chord weights were allowed to be non-negative real numbers see an example in Fig. 6.13, i.e., we implemented a fuzzy logic extension of the classical yes/no attributes of harmonic functions. The calculation method of these weights refers to embeddings of chords in minimal chains of thirds, details are provided in [118, Ch. 41.3]. The software then chooses an optimal path through the Riemann matrices of the composition's chords, i.e., for every chord an optimal function position its Riemann matrix is chosen. A HarmonRubette has recently been implemented for the Java RUBATO® by Ruhan Alpaydin, see [8].

6.4 Software-Base Experimental Music Theory: Rubato's MetroRubette for Brahms's *Sonata Op. 1*

Summary This section is a contribution of Guerino Mazzola, Ruhan Alpaydin, and Bill Heinze to the usage of software for experimental music theory in the analysis of classical European music. We apply Mazzola's mathematical method of inner metric analysis and its associated software to Brahms's *Piano Sonata in C major Op. 1*. We offer a short introduction to the complex rhythmic structures found in Brahms's later works and define Mazzola's process of inner metric analysis using the RUBATO® composer software. We then apply the algorithm to Brahms's sonata before finally drawing conclusions about Brahms's early development of rhythmic complexities, such as rhythmic dissonances.

6.4.1 Introduction

The music of Brahms has often been noted for its complex metric and rhythmic structures. Despite being labeled as a classicist in his lifetime, musicians as early as Schoenberg remark[4] on the intimacies of metric phrase structure in Brahms's music. These structures have been analyzed in the later works of Brahms where they are readily apparent. A thorough mathematical analysis of the metric and rhythmic structure of his early works has not been completed.

Inner Metric analysis examines the metric structure of a conventionally notated piece of music without regards to the bar lines within that piece of music. The quantification of each note's metric importance creates a hierarchy independent of the notated meter. This is the rhythmic structure, a global "atlas" of the onset collection O, covered by the "charts" of maximal local meters. The relation between a note's rhythmic context and its position within a measure, or outer metric structure, creates metric tension, when the two conflict, or coherence when the two align. Analysis of this relationship in movements of the early piano sonatas of Johannes Brahms can provide insight into a stylistic trait that helps define his later work. As Anja Volk (neé Fleischer) has noted, this methodology also adds precision and support to earlier writings on Brahms by Schoenberg, Frisch, Epstein, and others

[4]Arnold Schoenberg, in his famous essay on Brahms [147], writes: "Further, Brahms created asymmetrical and irregular meter constructions and employed polyrhythm throughout his career, providing interest and ambiguity, and taking this freedom farther than had Haydn and Mozart."

[29, Ch. 3]. Not only does it develop a scientific credibility, but it can be applied on a global scale to a piece of music.

Application of inner metric analysis by Volk of the works of Dufay, Ockegem, Wilbye, Morley, Händel, Bach, Mozart, Beethoven, Schubert, Schumann, and others is consistent with current studies into metric perception [33]. When the process is applied to the later music of Brahms, it reveals strong discrepancies between the inner metric structure and the notated meter of the piece.

This section applies Mazzola's mathematical model of inner metric analysis [118, Ch. 21] to the first movement of Brahms's piano sonata Op. 1. This process has traditionally been applied to compositions with the intent to confirm previous scholarship. This paper uses the algorithm to explore new hidden rhythmic hierarchies that are not obvious or even audible.

The cryptic nature of these mathematically comprehensive structures would preclude their interpretation in previous performances and analysis. Much like the many forms of a tone row in a serial piece may not be immediately audible, the revelation of these rhythmic hierarchies provides new structures for which to listen and deepen our understanding of the music. The audibility of musical structures is a psychological criterion that should not be conflated with their objective score-based existence. Our analysis is not an approach of psychological perception, but it refers to objective score data.

Our analysis demonstrates *the early presence of complex rhythmic processes* within Brahms's music. Recognizing the early existence of these constructions is crucial to understanding Brahms's lifelong development as a composer.

6.4.2 Inner Metric Analysis

6.4.2.1 The Method

Mazzola's method of inner metric analysis is a mathematical process, which is designed to consider the rhythmic relations of a piece of music globally. The process considers a *metric weight* for each note or chord. The weight is a "quantification of the note's metric importance within the piece" [23]. Examining the relative weights of different notes reveals metric hierarchies within the composition. The information generated specifically from the onsets of notes reveals a discrepancy between the meter implied by the bar lines and the meter perceived by the listener. When these meters coincide, a composer establishes *metric coherence* [29, Ch. 1.2], as opposed to metric incoherence or dissonance.

The analysis finds all chains of equally spaced onsets within a piece. These regular repetitions form *pulses*. A (set-theoretically) maximal series of uninterrupted regular pulses forms a *maximal local meter*. The number of pulses of the maximal meter m minus 1 is the *length* $k(m)$ *of the maximal local meter*. Each maximal local meter is assigned a weight determined by its length. Each note is then given its *metric weight* by considering the sum of the weights of all maximal local meters

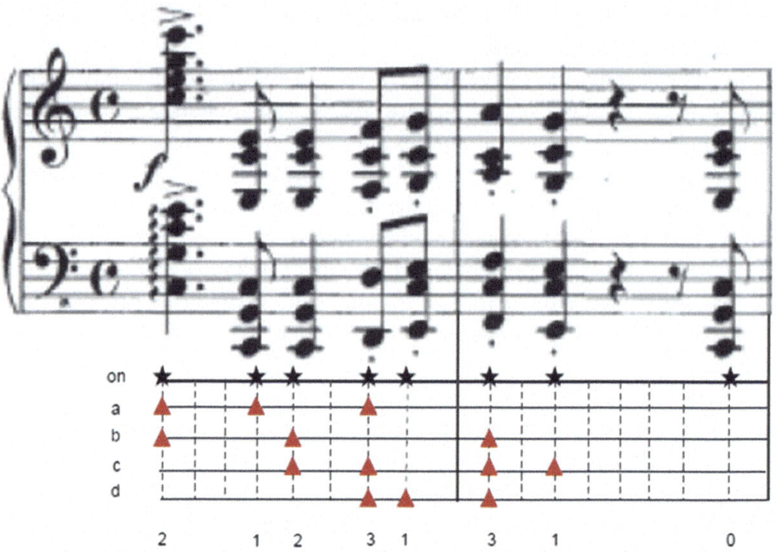

Fig. 6.14 Each onset is identified then each local meter of length 2 or more is labeled. Here we set $p = 0$ for the sake of simplicity. © Guerino Mazzola

which contain that note. For each note onset o of the set O of onsets in a given composition, the metric weight $W_{l,p}(o)$ sums the power function k^p of the length k of all maximal local meters m that have a minimum length l and contain o:

$$W_{l,p}(o) = \sum_{m,\ k(m) \geq l,\ o \in m} k(m)^p.$$

Figure 6.14 demonstrates the basic processes. Each note receives a weight based on the sum of the maximal local meters that contain it. In practice adjusting the value of p within the equation will place more value on the length of the local meters over the number of meters which contain a specific note. In this example, all local meters are treated equally regardless of length ($p = 0$), the weights are shown at the bottom.

We should stress that this method has an intrinsic musical meaning, it is not a purely mathematical formalism. In fact, if you choose a determined onset, a maximal local meter through that onset implies that the corresponding notes are participating in that local meter, which is an *elementary rhythmic event*. The length of that local meter describes the weight of this rhythmic event in the given composition. Adding powers of the lengths of the maximal local meters containing that event means to add up the presence of that onset with regard to all possible elementary rhythmic events. This is an essential *musical* information about the rhythmic position of those notes.

Our concept of rhythm refers to Hugo Riemann's concept of a "metrical quality" which is a weight function on onsets of notes [142, pp. 7–9] that is related to periodic properties of the score. His concept of rhythm involves weights, which Riemann calls the "pulse". Weights in rhythm theory are also generated in [95]. Our theory is based on the elementary local meters of onsets. A more general mathematically valid theory should be conceived in future research, including also harmonic, melodic, or dynamic structures. We refer to [118, Ch. 21] for a corresponding theoretical background. Theorists have previously used a theory of rhythmic weights to understand metric displacement in Brahms, see [37].

Inner Metric Analysis has been applied by Beran and Mazzola [118, Part XI] as well as by Volk [29] to investigate the relevance of such an analysis for shaping performance. These investigations support the conclusions:

- that, according to the cited Volk investigation, audiences prefer dynamic shaping of a piece informed by its metric analysis against both random shaping or shaping by metrics from other compositions,
- that well-known pianists shape their agogic accents according to the said metric analysis by Beran and Mazzola.

6.4.2.2 The New Software Implementation of the Theory

A software for the described inner metric analysis method has been first implemented in Objective C as a rubette component (the MetroRubette) in Mazzola's RUBATO® software for Apple computers, see [118, Ch. 41.1]. The present implementation of this component has been created by Ruhan Alpaydin. It runs as a rubette in the Java-based RUBATO® Composer environment and features a number of remarkable visual improvements of the old version, see Fig. 6.15. The first improvement is that the metric weight graphics are shown with a piano roll representation of the score which also distinguishes different voices by corresponding colors. The second improvement is the ability to display all of the local meters that contain a specific note by selecting it. Together with a flexible feature for selecting and visualizing any regular onset grid, this new implementation creates representations of the computational analysis that are more visually accessible. The grid mesh value n signifies a drawn barline every n quarter notes. E.g., if the grid mesh is $n = 4$, the large vertical lines represent a measure of $\frac{4}{4}$.

In our approach, we consider *all* local meters (the non-maximal ones being contained in maximal local meters) and their global configuration as a "metrical manifold," a global composition in the terminology of [118]. Moreover, the local meters that may emerge from metric weights are mostly not recognized by visual inspection only. They are local meters whose identity is a consequence of metric weights of their onsets. This is a higher conceptual analysis than straight visual inspection. *It cannot be discovered by straightforward human cognition, but results from the calculation of the comprising global configuration of all local meters.*

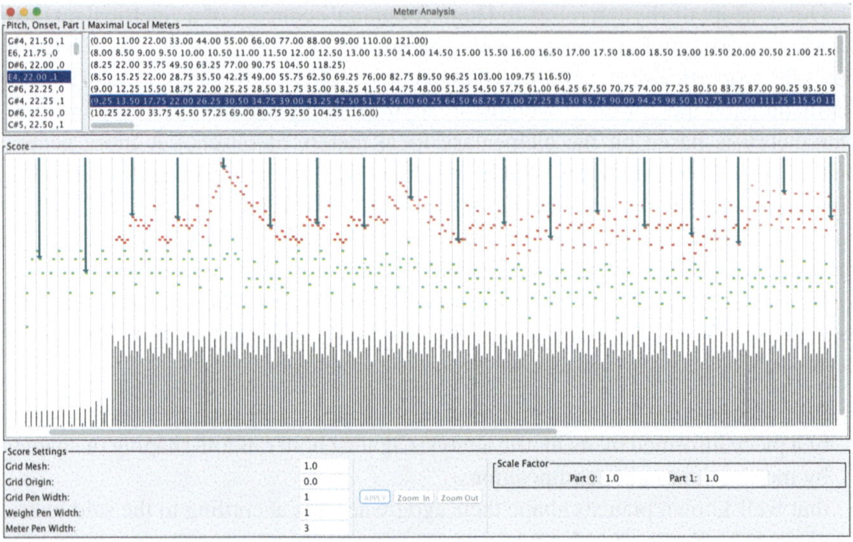

Fig. 6.15 The new MetroRubette displays metric weights for different voices (represented in different colors) and enables the display of single maximal local meters (selected and highlighted, shown on top) passing through selected onsets (selection top left). © Guerino Mazzola

6.4.2.3 Summary of Our Results

The analysis reveals the presence of metric dissonance in the sonata. These complexities most often exist in metrically closed sections. Each theme group in the exposition and recapitulation, as well as the development, have distinct metric sections visible in the analysis. Even though these sections often maintain the same implied meter, they frequently oppose one another and represent changes in texture. Further, we identify four types of metric dissonance in Brahms's early work: (1) metric shift, (2) metric competition, (3) metric modulation, and (4) dissonant hypermeters.

6.4.3 Inner Metric Analysis of Piano Sonata Op. 1, Movement 1, Allegro

Here, we apply inner metric analysis to the three major sections of the first movement of the sonata: Exposition, Development, and Recapitulation. We chose these sections, because they are not only formally distinct, but also rhythmically closed. The MIDI files for our analysis were derived from the score [17] by Bill Heinze. The inner metric analysis in this paper is performed with the variable $p = 2$ and $10 \leq l \leq 24$. The p value ensures that the length of local meters is appropriately taken into account. The restrictions on the value of l limit the importance of small runs of notes of shorter duration, and help elucidate the less obvious meters within the sonata. The specific l values are indicated along with the discussed examples.

6.4.3.1 Exposition mm. 1–89: First Theme Group mm. 1–38

The first thematic group of the sonata is notated squarely within Brahms's bar-lines. Hardly any notes are tied in such a way to obscure a notated downbeat. However, when inner metric analysis is applied, a hidden $\frac{5}{4}$ meter appears, see Fig. 6.16. The new meter is not expected, but sets the tone for the rest of the exposition. The sonata begins with a tension created by this incoherence, with two competing meters, one notated ($\frac{4}{4}$), and one composed ($\frac{5}{4}$). This meter has a remarkable length of 26 consecutive $\frac{5}{4}$ pulses (Fig. 6.17).

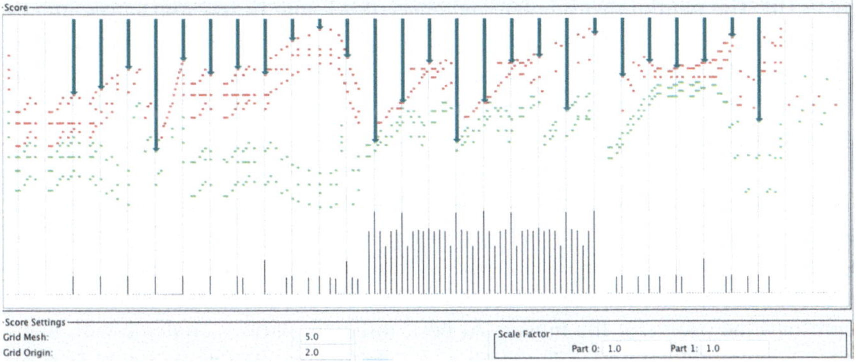

Fig. 6.16 Brahms Piano *Sonata Op. 1* mm. 1–38. The piano roll representation shows the pitches of the sonata. Here, $l = 22$. The long weight lines here (coinciding with the grid) show the piece as if it were in a quintuple meter. Note the 26 consecutive pulses. © Guerino Mazzola

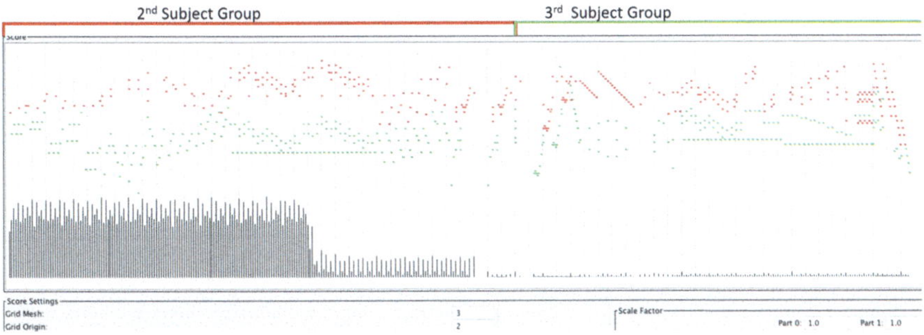

Fig. 6.17 Brahms Piano *Sonata Op. 1* mm. 39–88, the second and third subject groups. Here, $l = 22$. The anticipated quadruple meter is better reflected as a triple meter, as shown by the high weights on the $\frac{3}{4}$ grid. © Guerino Mazzola

6.4.3.2 Exposition mm. 1–89: Second Theme Group and Closing Theme Group mm. 39–88

When metric analysis is applied, the second thematic and closing group present a strong $\frac{3}{4}$ meter with a 2 bar phrase structure.The analysis reveals a hierarchy that one would expect of a triple meter: a strong downbeat with a slightly weaker second beat and an even weaker third beat. Like the first thematic group, inner metric analysis is incongruent with the notated meter of the music. More so, the triple meter disappears when the two hands are separated, see Fig. 6.18. The right hand maintains the repetition every six quarter-notes, but this could be heard as a one and a half bar hypermetric structure. This discrepancy between the original meter observed, the meter most likely perceived by a human audience (a clear $\frac{4}{4}$), and the meter observed when the hands are separated reveals a salient point on the music of Brahms: *even his earliest works contain metric ambiguity.* See also [30, p. 7] discussing competing meters.

Likewise, without the third subject group the global triple meter disappears, see Fig. 6.19. The pattern that emphasized every third note is apparent intermittently in the score. However the irregularity of the second group alone destabilizes the previously observed meter. The context of this meter raises some concerns. The sonata may be so structured that smaller local meters are largely unimportant to global shifts.

When looking at the exposition as a whole, see Fig. 6.20, the sonata seems to be divided into three distinct metric sections. These directly correspond to the sonata's harmonic and thematic structure. The first group is slightly irregular, the second is metrically dense and the third is sparse. This is consistent with Volk's analysis of Brahms's second symphony [30] viewed as a whole, no single meter accurately represents the profile of the music. At best, three bar phrases coincide with many of the most-weighted notes. These phrases, which contradict the standard four bar phrases of earlier composers may represent a young Brahms experimenting with hyper-meter. Likewise, the first theme group and closing theme group are the most

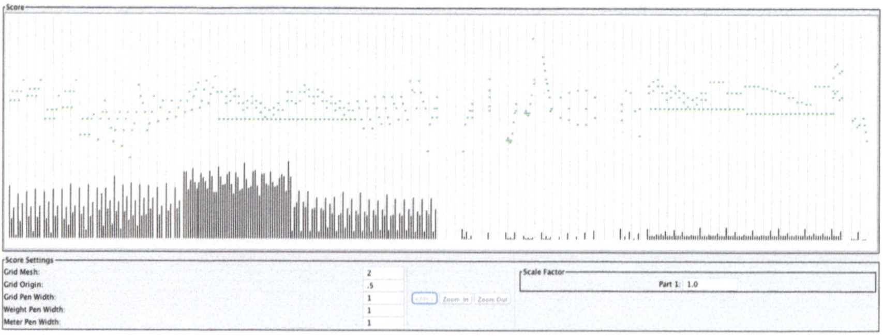

Fig. 6.18 Here, the weight of the right hand is dropped entirely, and $l = 12$. The triple meter found in the figure vanishes. © Guerino Mazzola

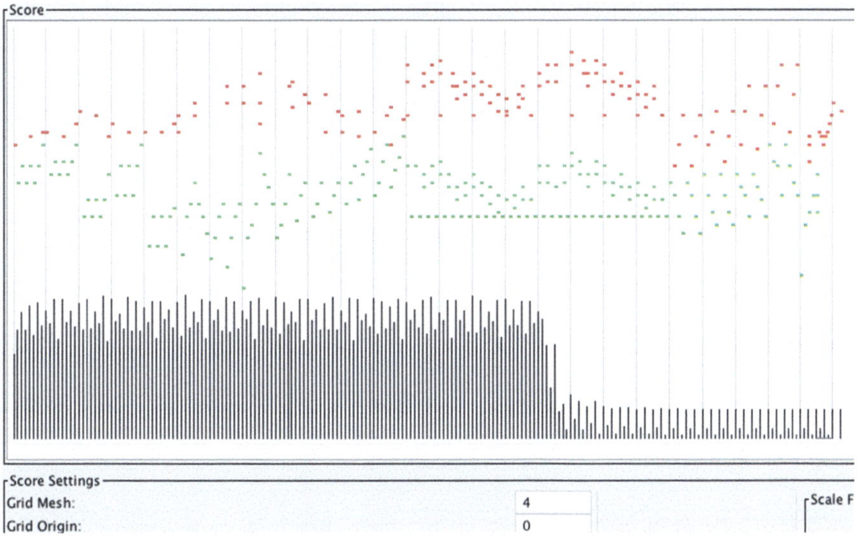

Fig. 6.19 Inner metric analysis reveals meters on local and global scales. We show mm. 39–67, the second group. Here, $l = 10$. By our analysis, this group without the third subject group lacks a clearly defined meter. © Guerino Mazzola

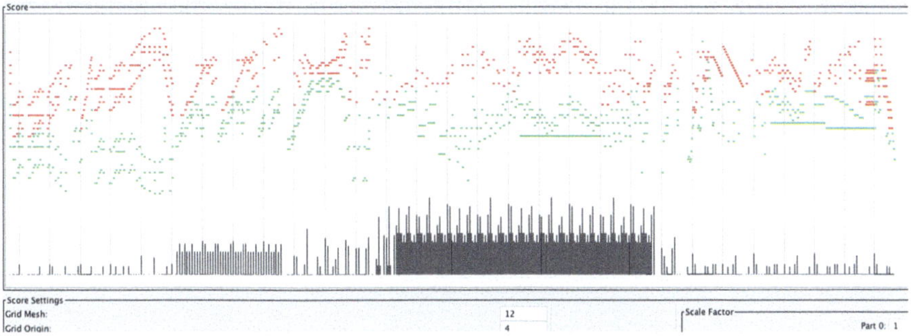

Fig. 6.20 Brahms Piano *Sonata Op. 1* mm. 1–88. Here, $l = 22$. Its exposition is parsed out into three bar hypermetric phrases. The rhythmically dense second thematic group serves as a clear divider between the first and third subject areas. © Guerino Mazzola

disjunct when viewed globally. The second group is metrically regular. This arc of metric stability will become important when we consider the recapitulation. It is also important to note that our inner metric analysis reveals a formal distinction between each of the subject groups. While form has traditionally been a problematic topic in music theory, this analysis reveals explicit metric partitions between formal sections (Fig. 6.21).

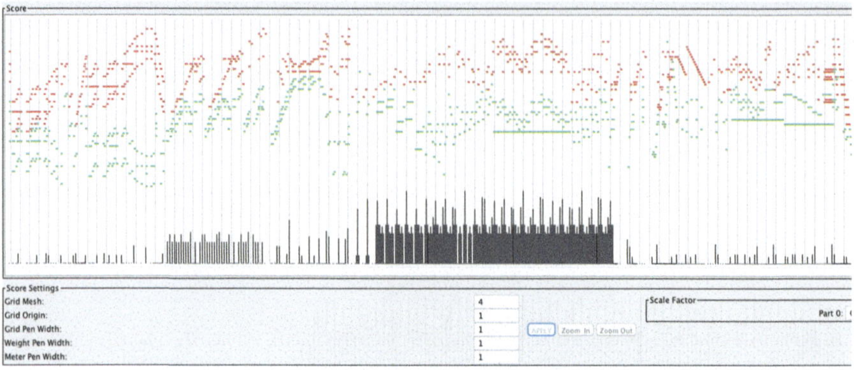

Fig. 6.21 Brahms Piano *Sonata Op. 1* mm. 1–88. Here, $l = 22$. The exposition is parsed out into the originally notated $\frac{4}{4}$ meter. Strong downbeats are common, but not consistent throughout the piece. © Guerino Mazzola

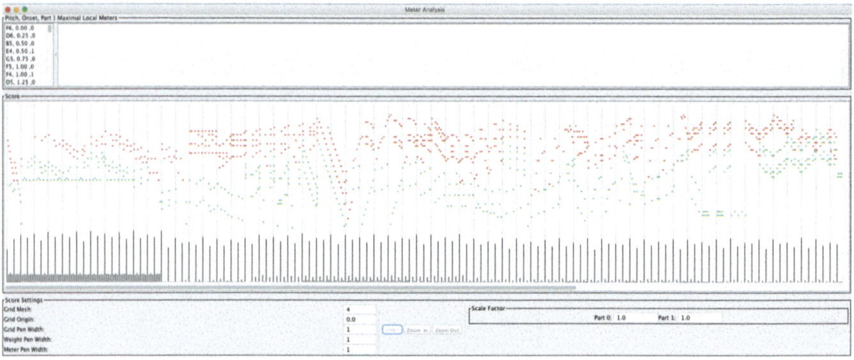

Fig. 6.22 Brahms Piano *Sonata Op. 1* mm. 88–149. Here, $l = 10$. The development has a very clear $\frac{4}{4}$ pulse, as shown by the grid lines. © Guerino Mazzola

6.4.3.3 Development mm. 88–172

The development always maintains a strong $\frac{2}{4}$ and even stronger $\frac{4}{4}$ metric period, see Fig. 6.22, where the grid is in $\frac{4}{4}$. While there are clear distinct sections where Brahms develops different material, the local meter remains the same. Even when broken into these sections, the meter is fairly regular.

While the meter of the development on a global scale is clear, Brahms creates this clarity through a competing tension. When examining the two hands separately, the right hand emphasizes what would be the third beat of every measure, see Fig. 6.23, while the left hand emphasizes the downbeat. This competition creates the clear hierarchy found in the original analysis, where the right hand creates the secondary displaced dissonant pulse, to the left hand's direct enforcement of the meter. This is consistent with the previous work done by Volk. Moreover, Mazzola in [111,

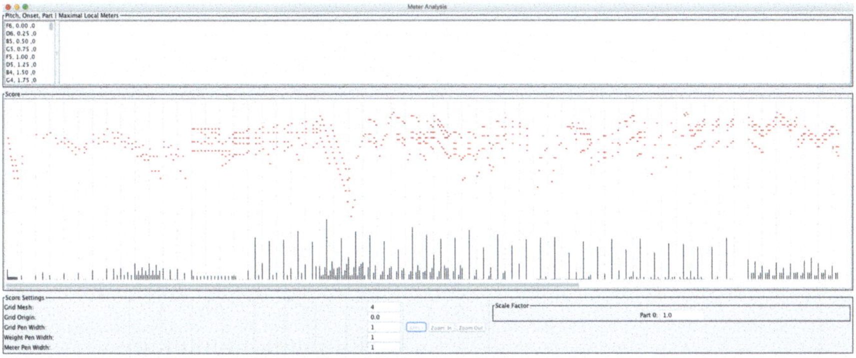

Fig. 6.23 Brahms Piano *Sonata Op. 1* mm. 88–172. Here, $l = 10$. The right hand establishes a competing down beat. © Guerino Mazzola

CH. 16.1] outlines Schumann's use of competing meters between two hands in his *Träumerei.* While scholars have long looked at both Brahms and Schumann's rhythmic complexity, the specific nature of the displacements was previously not rigorously mathematically understood. Brahms, as a student and long time friend of Schumann, likely picked up on his rhythmic language early in his career, as is evidenced by the displacement between the two hands in the development.

6.4.3.4 Recapitulation: First Theme Group

While the expository statement of the first theme group had a representative quintuple meter, the recapitulation abandons that meter. Instead, the first group returns in a triumphant stable four. At this point, rather than displace or obscure, Brahms reinforces. This reinforcement of the beat in the return helps create a sense of arrival in the music (Fig. 6.24).

6.4.3.5 Recapitulation: Second Theme Group to End

The second thematic group and the coda present no strong metric profile when considered together. While the incongruence to the notated meter is rectified in the first group, the second and third groups remain unstable. Notably, the third group contains no notes with heavily weighted onsets. This is understood to mean that the second group has many important notes without much order, while the third group has neither (Fig. 6.25).

When we consider the coda alone, we see that there is a standard quadruple meter, see Fig. 6.26. However, the hypermetric structure of the group does not match our expectations. What starts as an extremely regular four, becomes a clear set of three

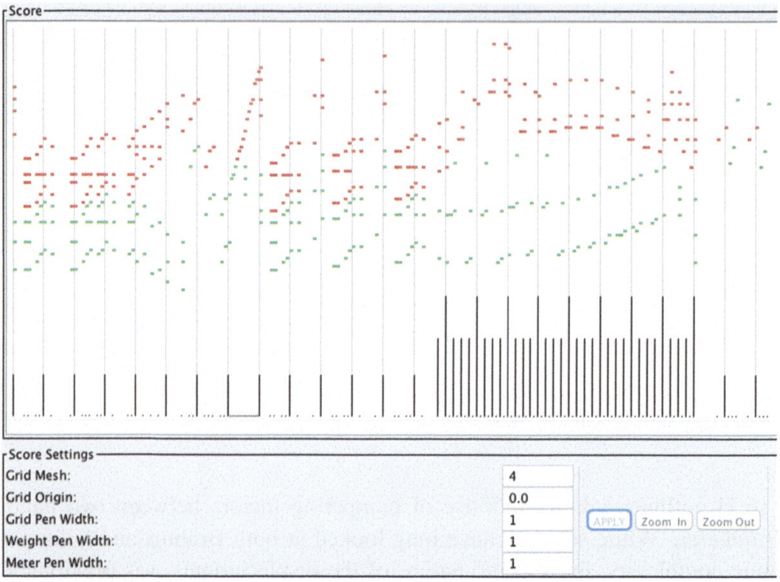

Fig. 6.24 The first group of the recapitulation. Here, $l = 22$. The beat emphasis is entirely regular. © Guerino Mazzola

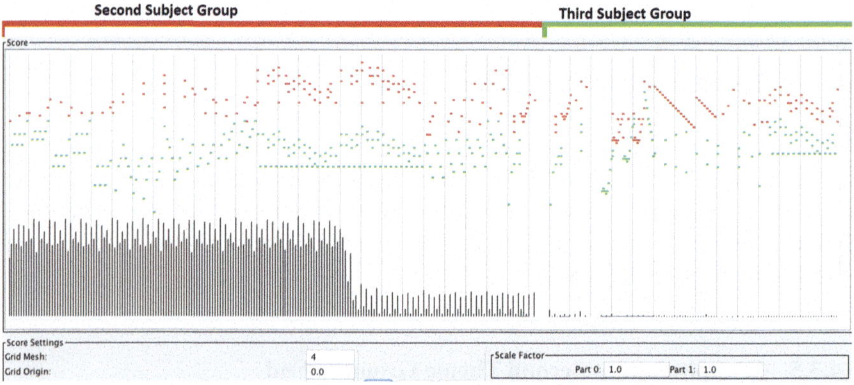

Fig. 6.25 The second group with the beginning of the third. Here, $l = 24$. Note the lack of clear metric structure at the beginning of the figure. © Guerino Mazzola

bar phrases with internal rhythmic weight symmetries. These structure is not evident from the score.

Summarizing the recapitulation, see Fig. 6.27, we can say that the first group is in a regular four, the second group largely irregular, and the closing section is sparse. Importantly, unlike the exposition, there is no larger hypermetric pattern across the recapitulation.

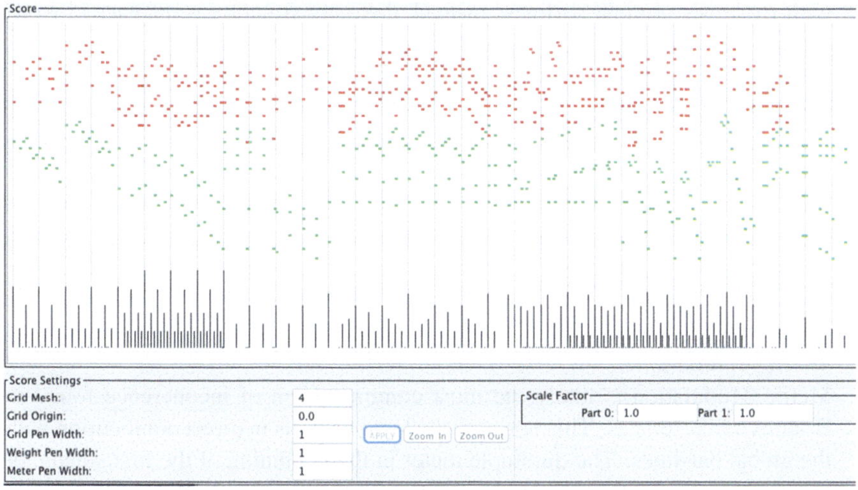

Fig. 6.26 The closing theme group of the recapitulation. Here, $l = 24$. © Guerino Mazzola

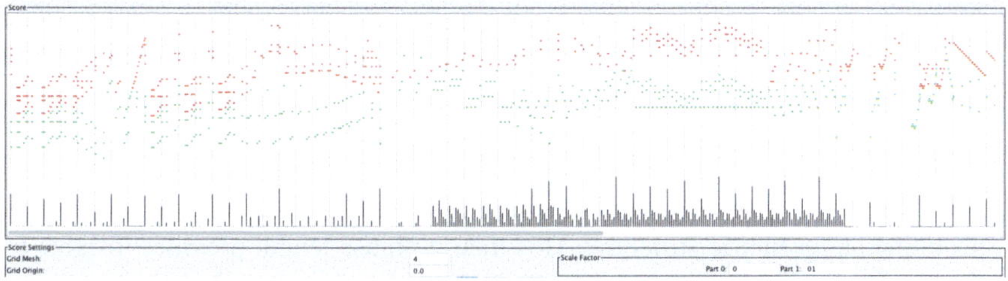

Fig. 6.27 A full selection of the recapitulation. Here, $l = 24$. © Guerino Mazzola

6.4.4 Conclusion

The incoherence of the local meter with the global structure occurs in early Brahms's as well as in his later symphonies. Rather than explain this metric incoherence as a technique that Brahms discovered at a later stage of creativity, it is important that we discuss it as a continuously evolving process in his compositional output. The analysis of the first sonata demonstrates the four following examples of rhythmic incoherence/dissonance found in Brahms's work. These events, or any combination thereof, contribute to the complex rhythmic language found in Brahms's music.

1. *Phase Shift*
   Phase shifts are instances where the outer metric structure and the inner metric structure are grouped in the same time signature, but the inner meter is displaced by some duration. When the exposition is analyzed as a whole, the algorithm displays a strong $\frac{4}{4}$ meter in the second group. The meter does not coincide

with Brahms's original barlines. Instead, the inner structure is displaced by one quarter-note. See Fig. 6.21. These shifts create an internal coherence. A listener with a score may not perceive the shift, but it is nonetheless irregular [30, p. 4].

2. *Metric Competition*
   Metric competition, which is found more frequently in the symphonies, is where two or more local meters exist in different voices. For example, the first piano sonata's second thematic group shows a strong meter that occurs every six quarter-notes in the right hand only. The left hand alone presents a simple quadruple meter. These two competing structures combine to create a third. Volk has observed how these competing structures in the symphonies of Brahms create irregularity [30, p. 7].
3. *Metric Modulation*
   Metric Modulation is likely the most common form of incoherence found in Brahms's first sonatas. This is where the local meter is in direct competition with the global bar-lines. The quintuple meter in the beginning of the first sonata, or the strong duple meter in the recapitulation of the third are both cases where the local meter is not represented in the bar-lines.
4. *Dissonant Hypermeters*
   Finally, even when there is a clear congruence between the notated bar lines and the results of the inner metric analysis, it is important that we look at the weight across multiple bars. For example, the right hand of the second thematic group in the exposition presents the algorithm with a $\frac{6}{4}$ meter that may be more realistically thought of as a series of one and a half bar phrases. Brahms is well known for his use of irregular phrase lengths [30]. Using inner metric analysis to examine larger metric structures This metric phrase length provides insights to how to phrase the music on a large scale. Hidden three, two and a half, or five bar metric structures could imply new interpretations of this oft-performed piece.

6.5 Performance

6.5.1 NotePerformer

NotePerformer [78] is an integrated sound library for Dorico, Finale, and Sibelius. This library adds a 1 s delay to playback so that it can analyze and interpret most markings in parts such as: dynamics, vibrato, articulations, extended techniques, and muting. This lets one attain a greater level of playback (and recording) than would normally be attainable from a music engraving software. It doesn't get the same level of playback that one can get from playing each part into a DAW individually, but it does take significantly less time and can be edited and exported directly from the engraving software instead of having to work between two different programs for a simple mock-up. It can play most Western musical styles and interpret most text based instructions. It has seven points of interpretation

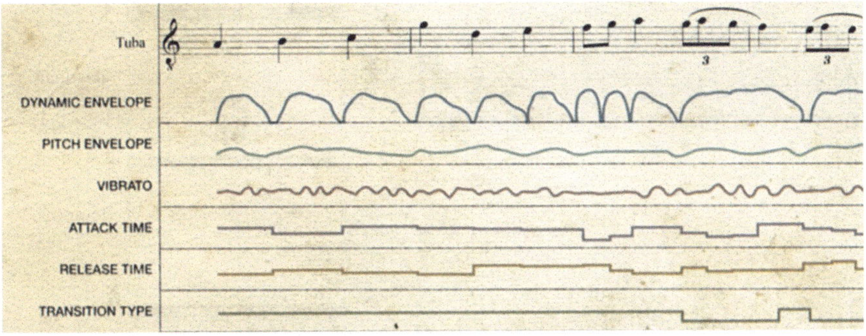

Fig. 6.28 An example of how NotePerformer analyzes music for playback

for sounds, dynamic envelope, pitch envelope, vibrato, timbre, attack time, release time, and transition type. An example can be found in Fig. 6.28.

In string parts it can simulate section sizes, with random variations and intonations between each individual player in the section. One can use half section and automatic *divisi* over chords. Different sounds are automatically assigned to different solo instruments and harmonics, and muting has the correct playback.

In brass parts one can get natural timbre changes from dynamics, one can use custom section building, they include five different mute varieties and accents, and *marcato* accents reach beyond the normally allocated maximum dynamic to create more of a brassy punch. One can use bowed mallet percussion instruments, and fast tremolos are corrected to a more natural speed. It has a wide variety of instrument types, including a selection of world instruments.

6.5.2 RUBATO®

As already explained in Sect. 6.3.2, the RUBATO® software was designed to create computer-generated performances from analytical input. One designated rubette, coined PerformanceRubette, was implemented for this task. Its input would be

1. the score to be performed, typically given in form of a MIDI file;
2. a number of analytical weights from the Metro-, Melo, or HarmoRubette.

With these data, the PerformanceRubette would provide the user with a number of performance operators that operate on the score, changing its tempo curve, dynamics, and articulation. Figure 6.29 shows the main theme of Bach's *Art of Fugue* that is being performed in four stages using its inverted melodic weight function. On top, we see the piano roll representation without performative deformation. Next, the inverted melodic weight is applied to the tempo. Then this performance is deformed in its articulation by the same inverted melodic weight, and last, this performance is deformed in its dynamics, again by the inverted melodic

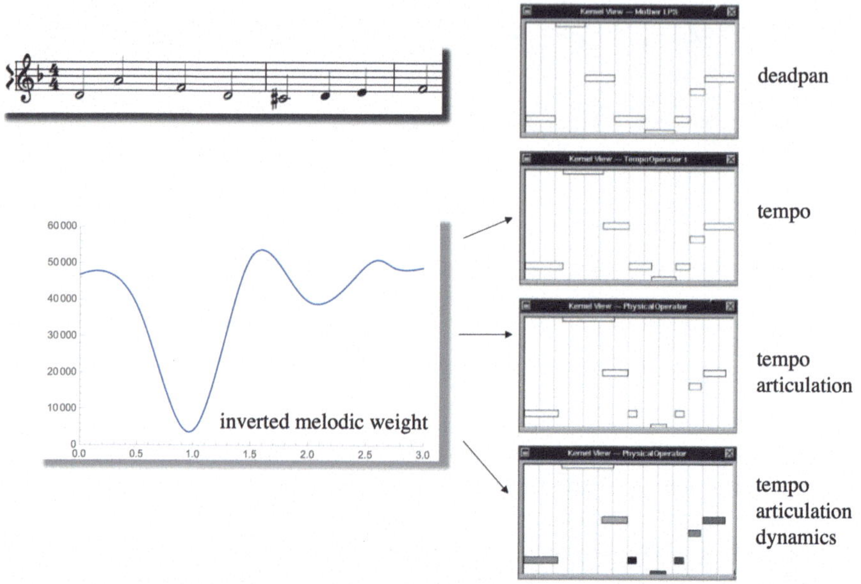

Fig. 6.29 The PerformanceRubette transforms the original main melody of Bach's *Art of Fugue* by a successive application of the melody's melodic weight to tempo, duration, and loudness. © Guerino Mazzola

weight. This scheme is called *stemmatic unfolding*. It proposes performance as being a succession of a sequence of intermediate performances, which are defined using analytical weights. It includes also splitting a score into specific parts, such as periods or left/right hand arts.

6.5.3 Melodyne

The software *Melodyne* [74] editor (Fig. 6.30) transforms audio data to note data and, after an unavoidable amount of editing, into MIDI data. Therefore, the performance research is open to a huge repertory of historical recordings.

6.6 Improvisation

The rise of software has led to two different directions in the branch of improvisation; one is software used as tools to help musicians improvise in real-time, such as Nodal, Max, SuperCollider, etc.; the other is software's autonomous impromptu accompaniment based on given direction.

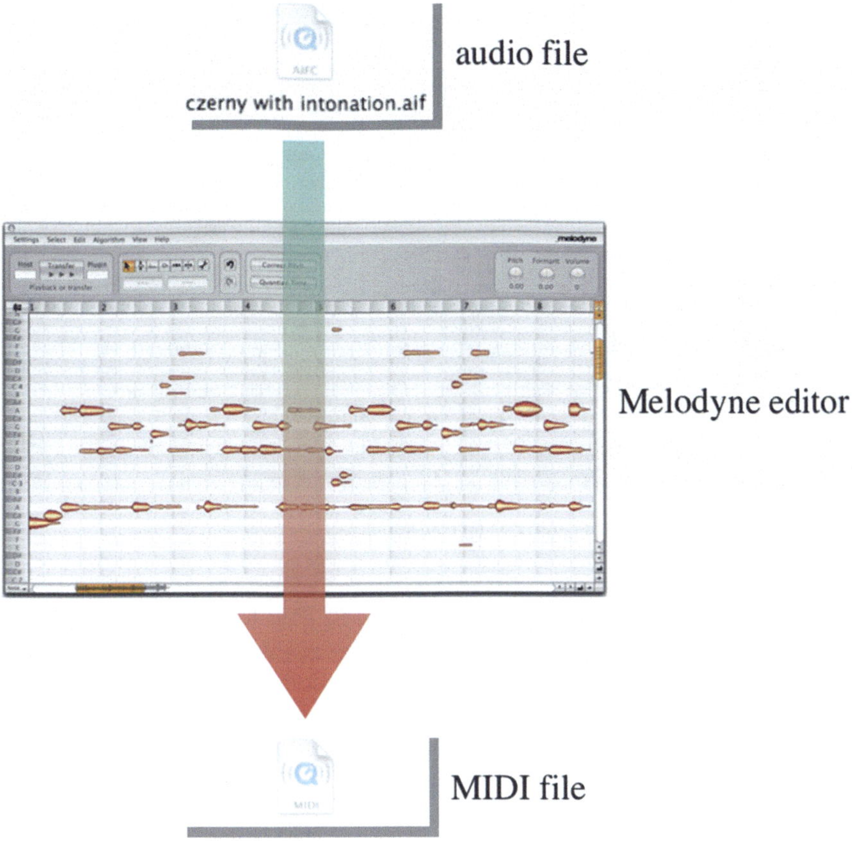

Fig. 6.30 The software Melodyne editor can transform polyphonic audio data into MIDI data and thereby opens research to performance analysis of historical recordings. © Guerino Mazzola

6.6.1 Nodal

In the case of *Nodal* [77], see Fig. 6.31, this is a custom-based software designed to assist musicians in improvisation. Its principle is very similar to visual programming. Users can create interactive maps by specifying nodes (musical events) and edges (connections between events). As shown in the figure, users can define number of virtual players of their choice, and the performance will start at the beginning of the node, along specified edges, to the end. The speed of playback is determined by the length of the edges. If a map is connected from start to end, the music forms a loop until a specified time or being manually stopped. Nodes can be, but are not limited to, effects, pitches, instruments, etc. By connecting with DAWs, users can use a MIDI keyboard for note input.

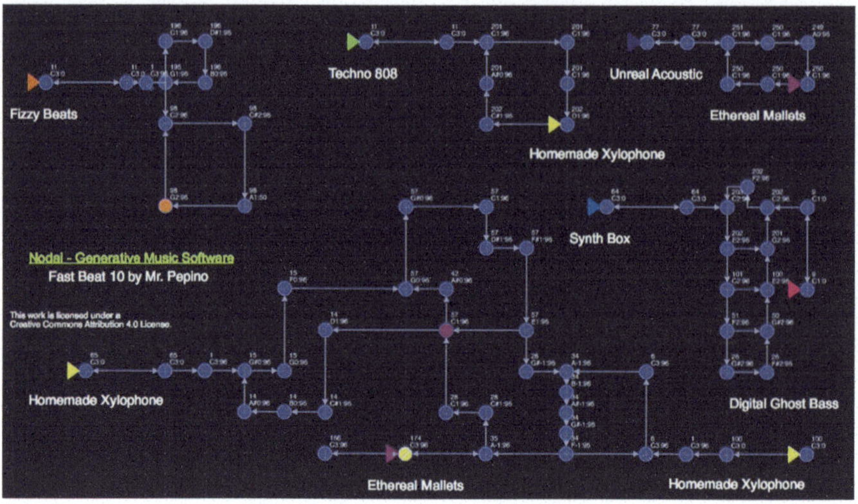

Fig. 6.31 FastBeat 10 by Mr. Pepino (Nodal Generative Music)

Like Nodal, Max and SuperCollider can also be used for improvisation. Users can use these applications to improvise through several means, e.g., using controllers (MIDI, gestural sensors, mobile phones, or desktop applications connected through Open Sound Control—OSC, an alternative language to MIDI, etc.), creating algorithms to automatically activate a patch of sound synthesis, using a live coding paradigm in which users are live-typing, etc.

6.6.2 Impro-Visor

From their definition [67]: "*Impro-Visor* is a music notation program designed to help jazz musicians compose and hear solos similar to ones that might be improvised. The objective is to improve understanding of solo construction and tune chord changes. There are other, secondary, things it can do, such as improvise on its own. It has also been used for transcription. Because rhythm-section (e.g., piano, bass, drums) accompaniment is automatically generated from chords, Impro-Visor can be used as a play-along device.(...) Lead sheets and solos can be constructed through either point-and-click or using a plain text editor (one is provided, but any editor can be used).(...) Solos and heads may be exported as MIDI files, with accompaniment if desired." Impro-Visor is free and runs on any platform that supports Java 1.8 or later, including: Windows, MacOSX, and Linux. See [67] for details. Figure 6.32 shows a typical notational interface.

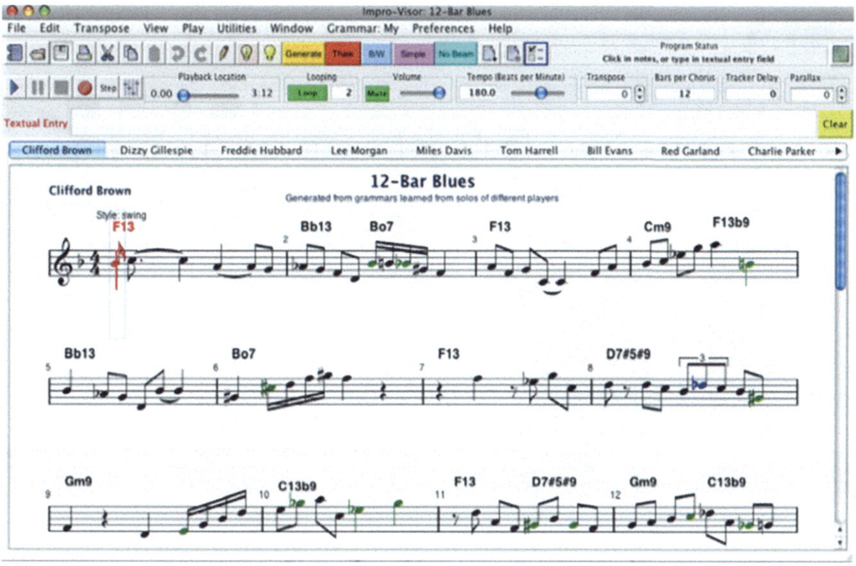

Fig. 6.32 An Impro-Visor window

6.6.3 *Band in a Box*

Band in a box [51] is a music accompaniment software package comes in handy when an existing melody seeks for quick accompaniment. With input like chord progressions and music style desired, it automatically generates a complete professional-quality arrangement of piano, bass, drums, guitar, and strings or horns. Unlike musicians improvisation, this kind of accompaniment is based upon a massive data analysis of existing songs and music. It is now trying to add more real tracks played by musicians, but how those tracks are used is still decided by an algorithm. The benefit of having this kind of software is to use it as a quick introduction to specific genre improvisation. However, the limitation is obvious too: it will restrict creativity. Figure 6.33 shows a typical working interface of Band in a box.

6.7 Notation

Now technology has promoted the electronicization of traditional music scores. For example, *Sibelius* [86], *Finale* [63], *Dorico* [59] and other softwares can support playback and graphically interactive editing of scores, or through MIDI files or MusicXML files. Even if there are many features yet to be developed, they

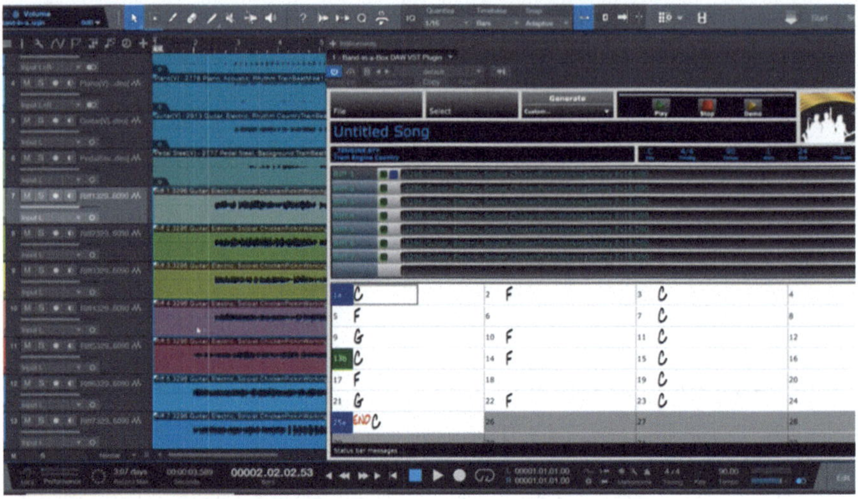

Fig. 6.33 Working interface of Band in a box

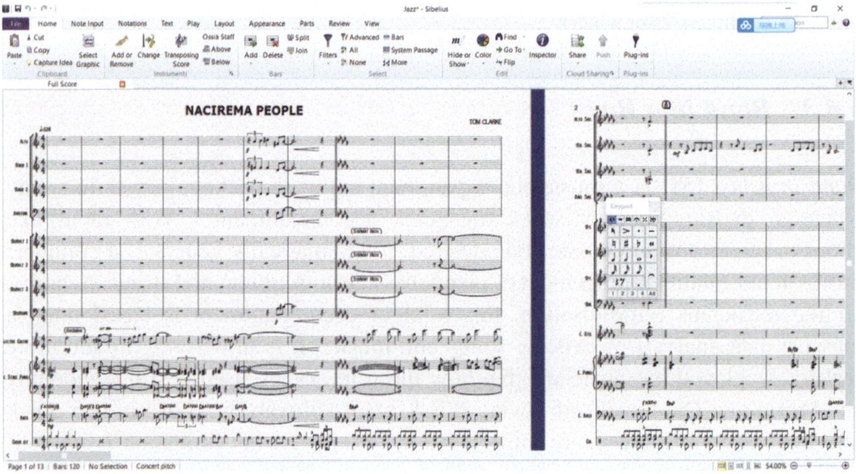

Fig. 6.34 Sibelius notation software

have become vastly popular because of their versatile functions and accessibility. Figure 6.34 is an example of a typical notation software's interface.

In the past, cultivating a composer was expensive. Composers from early eras needed to be able to imagine how their compositions would sound played by instruments in order to create a score. This extraordinary imagination usually came from the experience of being exposed to live music. Not everybody has access to trained musicians. But with software, working with virtual instruments, such as NotePerfomer, composers can hear a playback of their work. Because the software does not need to rely on the real-time performance of the band, it also gives the composer more independence. The editing process becomes more convenient and flexible. There is no need to go through time-consuming and costly steps of assembling bands, rehearsing, and re-scoring.

Despite the great convenience of the software, the excessive dependence will affect the creativity of a composer. For example, by default, the initial time signature of the editing interface is 4/4. If a composer can't think outside of the box, it is very likely that one will write under the default settings. However, a composer should keep in mind that the playback from the software is not 100% accurate. It is always better to work with real musicians if there is opportunity.

6.8 Education

6.8.1 Auralia

There has been numerous softwares launched for ear training purpose and *Auralia* [50] is one of them. It starts with the basics like pitch, intervals, chords, scales, tuning and rhythm, then progresses through cadences, rhythm dictation, harmony, jazz progressions and melodic transcription. Its advantages are self-evident; reducing the dependence on instruments and others during training, easy access, record performance for future reference. Its highly customizable options also prepare it for wide range of users. Its shortcomings are equally obvious; the sound of electronic synthesis has not been able to sound the same as the acoustic sound. The lack of certain overtones makes such training easier to identify in comparison to acoustic training. In order to be more comprehensive, musicians are recommended to have acoustical ear training in addition to Auralia. In Fig. 6.35, shows the content library of Auralia.

6.8.2 Syntorial

Today, composers are no longer satisfied with just the sound of using existing instruments. Mothership by Mason Bates is a good example of combining synths sound with orchestra. The invention of the computer gave composers enough freedom to expand more sounds, so comes the invention of synthesizers. For those who are not so familiar with the physics of sound, there are certain obstacles when learning synthesizers, so comes the invention of *Syntorial*. Syntorial is video game-like training software, that will teach people how to program synth patches by ear [89]. With almost 200 lessons, combined video demonstrations with interactive challenges, Syntorial provides hands on a programming aid on a built-in soft synth.

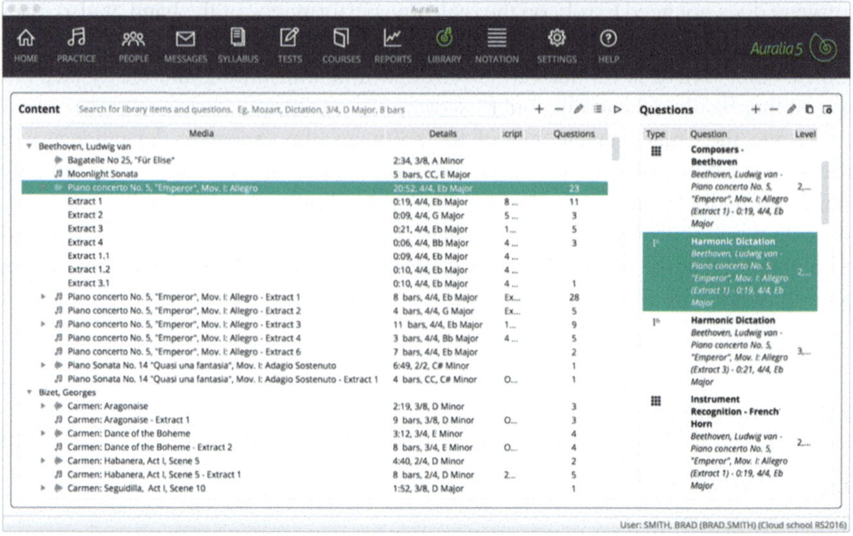

Fig. 6.35 The Auralia content library

Fig. 6.36 Challenge interface with content on the left

Figure 6.36 shows a typical challenge interface of Syntorial. The only downside of this software is that, in general, synthesizers have different interfaces. Composers may feel uncomfortable using other synthesizers after a long use of Syntorial.

Fig. 6.37 A Counterpointer score with error highlights in red

6.8.3 Counterpointer

Counterpointer [56] is a software for composition and education in counterpoint. From its definition: "Counterpointer is the first counterpoint instruction software ever to offer evaluation of both species and free counterpoint, and can serve both as a self-contained introduction to counterpoint and as a tool to accompany counterpoint textbooks.(. . .) Compositions can be saved, heard, printed, and exported as MIDI files." Figure 6.37 shows a score with error highlights in red, which are explained in specific windows.

6.9 Hardware

Summary Hardware for music is increasingly powerful for storage and processing purposes. This chapter is meant as a guide into hardware and its uses for musical composition and performance.

6.9.1 Input

Hybrid percussion studio pads are controllers for software systems you can pair with DAW's to increase efficiency for recording, composing, and live performance. Many of them offer basic touchpad use which you can program to create certain

Fig. 6.38 The NI Maschine

sounds, effects, or use in conjunction with add-on programs such as Max/MSP to create very unique effects.

NI Maschine [76] The Maschine by Native Instruments is a tool that integrates itself into many standard DAW's (see chap. 6.1) in order to simplify and facilitate composition and performance, see Fig. 6.38. With this tool you can produce beats, melodies, and harmonies by using a combination of touch sensitive pads, control knobs, and the virtual interface on the board itself. It allows you to browse sounds more quickly by using a control knob and one of the screens built into the board. It has a built-in variation generator that creates new melodies and drum patterns on its own with transitions.

Ableton Push [48] Like the NI Maschine, the Push is a tool that integrates itself into many DAW's but with a focus on Ableton Live, see Fig. 6.39.

6.9.2 MIDI Keyboards

When using a DAW or music notation software you have the option of using a point and click method or using your QWERTY keyboard[5] to input notes, durations, and other markings. When you combine this with a standard MIDI keyboard you can do this in a more precise manner. MIDI keyboards make it possible to choose octaves

[5]WERTY is the acronym that commonly describes today's standard keyboard layout on English-language computers. The QWERTY layout was patented in 1874 by Christopher Sholes, a newspaper editor and the inventor of the typewriter.

Fig. 6.39 The Ableton Push

Fig. 6.40 The CME Xkey

when inputting notes, add chords quickly, and you have the option of live recording notation, which when paired with adequate quantization will produce very accurate results, allowing you to input notes in a much faster manner than using point and click or QWERTY keyboards. For basic solutions you can use something such as the *CME Xkey* [55] (Fig. 6.40), which is also a highly portable option with basic octave, pitch bend, and modulation options.

For more advanced solutions you can use something similar to the *Roland 88-key FA-08* model [85] with weighted keys, touch pads, and many minute adjustment controls, see Fig. 6.41.

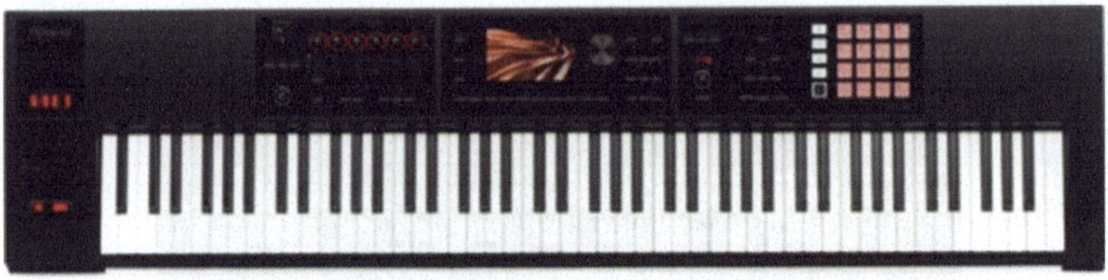

Fig. 6.41 The Roland 88-key FA-08 model

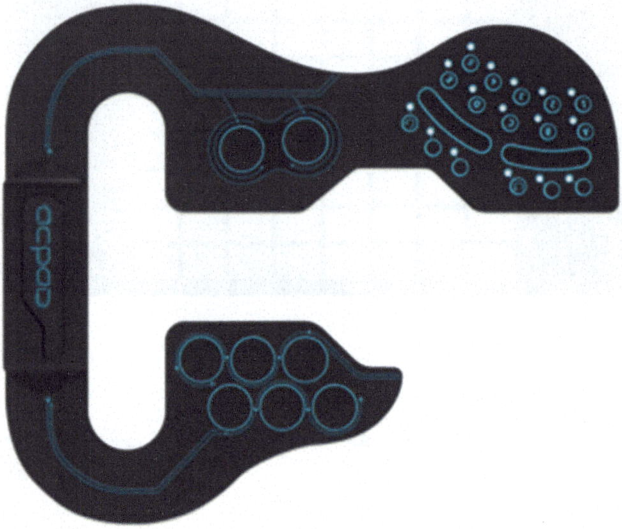

Fig. 6.42 The ACPAD

6.9.3 Alternative MIDI Input Tools

ACPAD Before ceasing production *ACPAD* (see Fig. 6.42) was the first Wireless MIDI Controller for Acoustic Guitar [49]. You can mount it to an acoustic guitar to transform guitar performances with MIDI input, this allows for some new and interesting sound combinations and self accompanied performance built into the attachable interface.

EWI Akai's *EWI* (Electronic Wind Instrument) [62] (see Fig. 6.43) is a MIDI input that allows for the use of air to enhance sound production characteristics, this tool allows the user to choose which instrumental fingerings to use and then apply wind and input MIDI through playback.

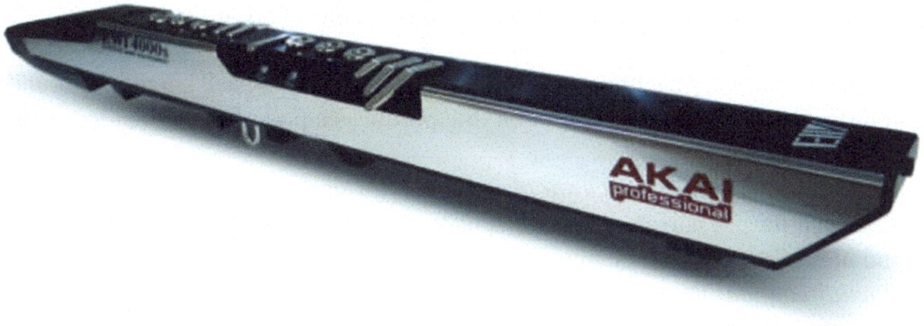

Fig. 6.43 Akai's EWI

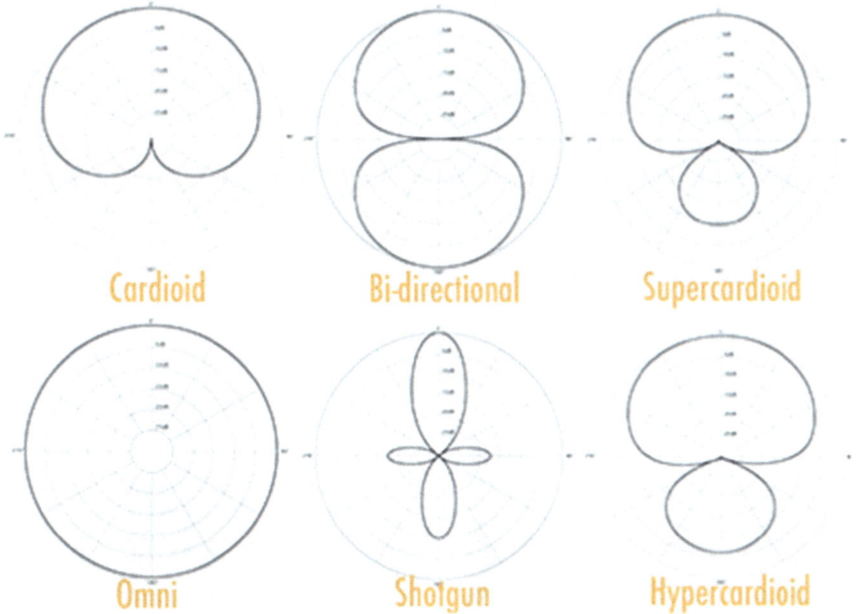

Fig. 6.44 Microphone types

6.9.4 Microphones

There are many different types of microphones available for use. Below we will explain each kind and their uses. We will discuss polar patterns, diaphragm sizes, and finally microphone type, see also Fig. 6.44.

1. *Polar patterns* describe a microphones field of sensitivity and how each type picks up sound. Below are the most common types of polar patterns.
2. *Cardioid microphones* capture sound from the front and ignore everything emanating from the rear and sides. This front-focused capture method allows

you to point directly at a sound source and isolate it from unwanted ambient sound.

3. *Bi-directional mics* capture sound from both the front and back but rejects sounds coming from the sides. This type of mic produces a polar pattern that looks like a "figure-8."
4. *Supercardioid and Hypercardioid microphones* are more directional than cardioid and have a narrower area of sensitivity. They provide improved isolation and higher resistance to feedback.
5. *Omnidirectional microphones* equally capture sounds from all directions. This type of microphone has a more natural sound quality because accept a room's sound as is.
6. *Shotgun mics* are a more aggressive form of hypercardioids. They feature a tube like design that has a longer pickup range. They are used as overhead mics for capturing sounds in a room. However, shotgun mics often have very poor sound quality.

Multiple pattern microphones allow you to switch between different polar patterns. Many of today's USB condenser microphones have a feature that provides the option to select a polar pattern. Other types provide the same flexibility through changing the mic head (Fig. 6.45).

Microphones pick up sounds through their diaphragm, a thin material that vibrates when it comes into contact with sound. The size of the diaphragm affects the microphone's sound pressure level handling, sensitivity, dynamic range, and

Fig. 6.45 Diaphragm sizes from left to right, medium, small, large

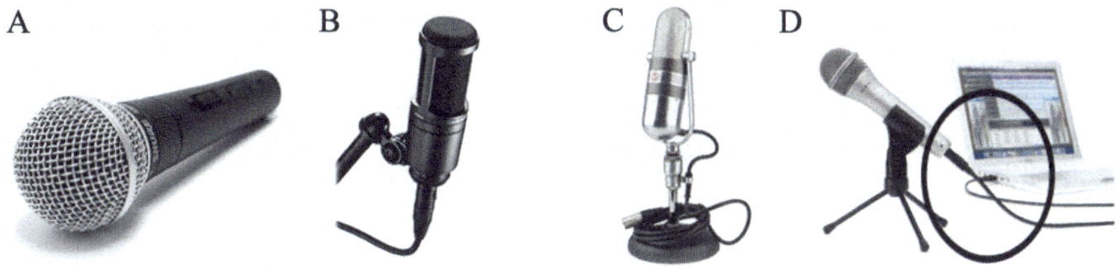

Fig. 6.46 Four microphone types. (**a**) Dynamic. (**b**) Condenser. (**c**) Ribbon. (**d**) USB

internal noise level. There are three standard mic diaphragms classifications, small, medium, and large.

1. *Small* Diaphragm mics are lighter, easier to position, and ultra-responsive. These types of mics are great for capturing acoustic guitars, hi-hats, cymbals, and other instruments with sharp transients.
2. *Medium* Diaphragm microphones combine the characteristics of small and large diaphragms. They tend to have a slightly fuller and warm sound similar to large diaphragms while retaining some of the high-frequency content that small diaphragms capture.
3. *Large* diaphragm microphones are often used to capture everything from vocals to all types of instruments, room spaces, and more. The larger diaphragm is capable of capturing sounds in great detail in order to produce a high fidelity recording.

There are four types of microphones used for recording music, see Fig. 6.46. They use either XLR or USB cables for connectivity. These microphone types are Dynamic (a), Condenser (b), Ribbon (c), and USB (d) microphones.

1. *Dynamic microphones* are the cheapest of the four. They take in sound through audio signals generated by a moving coil within a magnetic field which makes this type of microphone less sensitive to the pressure levels of sounds and high frequencies.
2. *Condenser microphones* are the most responsive of the microphone types. They have a thin conductive diaphragm attached very close to a metal backplate. This configuration works like a capacitor where sound pressure vibrates the diaphragm which creates an electrical charge to produce the audio signal output. These microphones use capacitors instead of actual moving coils. Condenser microphones generally require the use of an external power supply, internal batteries, or phantom power supplied by a mixer.
3. *Ribbon microphones* have a thin metal ribbon instead of a diaphragm allowing them to pick up the velocity of the air instead of just air's relative displacement. This design makes them more sensitive to higher frequencies.
4. *USB Microphones* are becoming more popular moving forward for their immense adaptability to the recording situation. Their design can resemble any of the aforementioned microphone types, and can replace most microphone types for

a large number of projects. They include an onboard preamp and an analog-to-digital converter. The preamp eliminates the need to be connected to a mixer or external mic preamp while the analog-to-digital converter changes the mic's output from analog (voltage) to digital (data), so it can be plugged directly into a computer. This ability makes USB mics ideal for AW software.

6.9.5 Output

Soundboards and audio converters connect your computer software to external speakers, there are many options as to how this connection is made such as USB, Auxiliary cable, stereo cables, and optical digital audio cables. Soundboards help to alleviate the limitations of computer sound cards as well as give you an avenue for more finely tuned control of your outputs, see Figs. 6.47 and 6.48.

A quick note on Computer hardware itself. As computers continue to advance we must account for the increased system requirements to run the software

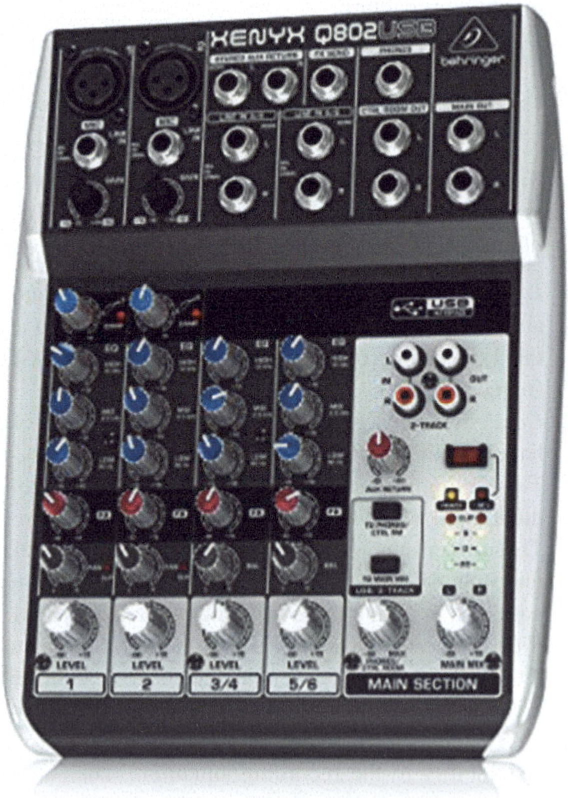

Fig. 6.47 The Behringer Xenyx Q802USB is an example of a simple solution for better control of recording and playback

Fig. 6.48 Something like the Behringer UltraGain Pro-8 ADA 8000 can be used for simple solutions as well as to create an ambisonic environment

described in this chapter. A 4500 dollar desktop from the 1990s, for instance, with 8 MB of RAM, less than 250 MB of hard drive space, and a processor whose frequency is 25 MHz wouldn't even compete with a 200 dollar laptop today, let alone be able to run any modern software. Today, workstations can have countless terabytes of Hard drive space, hundreds of gigabytes of RAM tuned hundreds of times faster than the previously listed computer. The CPU's of today run up to 5 GHz on consumer models with over 16 cores instead of single core processors, see also the following table. Here is the meaning of the symbols in this table: MB = MegaByte, GB = GigaByte, TB = TeraByte, MHz = MegaHertz, Mega roughly 10^6, Giga roughly 10^9, Tera roughly 10^{12}, RPM = Rotations Per Minute, IDE = Integrated Drive Electronics, SATA = Serial Advanced Technology Attachment, SSD = Solid State Drive, HDD = Hard Disk Drive, PCI = Peripheral Component Interconnect, PCIE = Peripheral Component Interconnect Express, CRT = Cathode Ray Tube, AGP = Accelerated Graphics Port made specifically for Graphics Cards, ROM = Read Only Memory, RAM = Random Access Memory.

Part	1995 PC	2019 PC
Memory	8 MB at \$ 400 per 4 MB at around 33 MHz	8 (GB) in common with the possibility of 256 GB, which is 8000–256,000 MB \$ 30 per 8 GB stick within 2133–4800 MHz
Hard drive	400–1000 MB 5400 RPM IDE format 800 kB/s (0.8 MB/s)	Evolved to SATA III SSD's, M.2 SSD's, and massive HDD with storage option up to 8 TB 7200RPM → 50 MB/s, SATA SSD → 450 MB/s, M.2 SSD up to 2500 MB/s
Processor	33 MHz, single core/single thread	4000 MHz (+) up to 32 cores/64 threads with countless optimizations
Video	24 Bit accelerated	PCI Express 3.0 (replacing PCI 1 and 2, then AGP) with 6 GB of dedicated RAM, for about \$200
Monitor	14″ CRT 800 × 600 resolution	24″ plus wide screen LCD/LED 1920 × 1080–4096 × 2180 resolution
Sound	Sound blaster 16 (16 Bit)	2 Bit, PCI Express, 3d, quad core processors with onboard RAM
Modem	28–54 kB/s internet speeds	Obsolete, except for users in very rural areas, replaced with fiber, cable, and wireless internet, internet speeds up to and surpassing 1 GB/s
Optical disk	2× CD-ROM	BlueRay, DVD, some CD-ROM, all but rendered useless due to online streaming

Chapter 7
New Concepts of Musical Instruments

7.1 The Classification of Instruments

Before we expand on the realization of new musical instruments, we must define their categorical rumination. Sound is generated through the input of force on matter that then oscillates and transfers energy which passes in the form of waves through the air. The preeminent distinction in current instruments stems from the differences in the input.

7.1.1 Acoustic/Mechanical

An established classification of instruments is the Hornborstel-Sachs system. It divides instruments into five main categories, where the first four are acoustic instruments, the fifth being the electroacoustic type. Here are the acoustic types:

- Idiophones—Instruments which produce sound when the entire body of the instrument vibrates. These include mallet percussion instruments, claves, cymbals, and triangles.
- Membranophones—Instruments with a stretched membrane that is responsible for producing the sound. This includes any drum.
- Chordophones—Instruments where sound is produced by the vibration of any number of strings stretched between two points. This includes pianos, members of the violin family, and guitars.
- Aerophones—Instruments where sound is produced by vibrating air within the instrument. These include woodwind and brass instruments.

G. Mazzola et al., *The Future of Music*,
https://doi.org/10.1007/978-3-030-39709-8_7

7.1.2 *Electroacoustic/Electromagnetic*

These are instruments which have electromagnetic action, or where sound is produced by electromagnetic means. This includes pipe organs with electric air valves, electric guitars, theremins, and synthesizers.

The software-driven electroacoustic instruments are split into two subcategories:

- Instruments, which generate the sound wave types following basic synthesis methods. These include the classical Fourier synthesis (overtones), frequency modulation, sampling and granular synthesis. Sampling technology refers to recording pitches under different techniques and dynamics. These audio fragments are associated with certain algorithms and continuous controllers, so that the user's specific operation can trigger a certain playback. In other words, it is a pre-recorded audio or a combination of audio fragments for a virtual instrument simulation technique. Hence the result of simulation will be greatly limited by the quality of sampling, such as MIDI velocity stratification, techniques, position in string instruments.
- Instruments which are modeled as software constructs, where the physical output waves are generated by the simulated movements of such software instruments. These instruments are called physical modeling. Physical modeling refers to the simulation of the sound of an instrument by models derived from a mathematical formula that represents the physical sound production architecture of the instrument. Due to the universality of the mathematical formula, the final product is no longer restricted by physical stratification. One may, for example, model the vibration of a string of glass, 3 m of thickness and 2 km of length.

  One advantage of physical modeling is that it reduces the occupation of data space. Compared with sampling instruments which can take up hundreds of gigabytes, physical modeling relies more on the calculation of CPU. Physical modeling instruments can change parameters in real time from the beginning of a sound to the end, for example the breath control of a wind instrument, or the bow pressure of a bowed instrument. This can also be a drawback as the sound is often too precise to be real. The physical modeling also does not eliminate the necessity to learn to play it, the musician is not eliminated by this technology.

7.2 Expansive Realization

In Expansive Realization, the creator aims to modify a current design in order for the original instrument to reach a given goal set by new parameters. The variances of parameters can be based on a singular or combined concepts. In Fig. 7.1, we show the "Double Horn, Double Bell", an instrument that addresses timbre, technical efficiency, volume, and many other aspects that comprise the identity of an instrument.

Fig. 7.1 Christine Chapman with the "Double Horn, Double Bell". Image reprinted with kind permission from Ensemble Musikfabrik/Valerij Lisac

The MIDI language has proven itself to be integral to the conception of modern music in the twenty-first century. MIDI (Musical Instrument Digital Interface) is a protocol designed in order to communicate data between electro-acoustic instruments; see our discussion of MIDI in Sect. 6.1.4. So MIDI is not a proper instrument, but a crucial communication device for electroacoustic instruments.

The concept of expansion applies to the language of MIDI as well. The success of the MIDI 1.0 implementation since 1983 is well established in its popularity, but as if it were a natural phenomenon, the question, "can it be improved" came into existence. It resulted in the announcement of MIDI 2.0 prototype in 2019 by the MIDI Manufacturers Association (MMA) and the Association of Music Electronics Industry (AMEI).

Traditionally MIDI has 7 bits of coded information, or resolution. These 7 bits create the parameters that define the range disparity as 27 or 128 different units. For example, loudness, termed velocity, can be given a value of 0 through 127, but MIDI 2.0 has deemed it necessary to expand its resolution to 32 bits. Through MIDI 2.0 that same value of velocity can now be defined from a range of 0 through $2^{32} - 1 = 4,294,967,295$.

Extensions of acoustic instruments are frequent even for the most standardized piano. Boesendorfer has extended its grand piano types to include the "Boesendorfer Imperial", where the keys are extended by nine keys in the low range.

To play quartertone compositions by Alois Haba, August Förster invented the quartertone grand piano in 1923, see Fig. 7.2. There are recordings of quartertone piano music: Ivan Wyschnegradsky: 24 Preludes in quarter-tone system, 1924 [94].

Fig. 7.2 Ivan Wyschnegradsky at his quartertone piano. © Photo René BLOCK—Ivan Wyschgnegradsky 1979

7.3 Creative Realization

The creation of a new instrument is a goal-oriented endeavor, that is to say, it serves a purpose in order to reach a desired result. But there are instances in which sounds needed cannot be produced by expanding the capabilities of traditional instruments as discussed in Sect. 7.2.

There is a commonality in methods used to create new instruments, which means that most instruments share certain fundamental characteristics in the production of air waves. In the creative realization of a sound one doesn't change the fundamental way in which sound is produced, but one creates different means of communications through matter resulting in completely new sounds.

Creative Realization is dependent on having a general understanding of the parts of a musical instrument and their relationships towards one another. We will discuss this topic from the point of view of what an instrument is, namely the interface between musician's input (performer or composer) and sound output. So three things can be created: new input, new interface, new sound output, see Fig. 7.3.

1. New input (Gestural Communication)

   - This can be new symbols (score types, software data, ideas, sound imagination)
   - New gestures (how to access the interface, hands, lips, body, electronic tools)
   - New emotional input

2. New interfaces

   - Gestural access modes
   - Data communication (MIDI etc.)

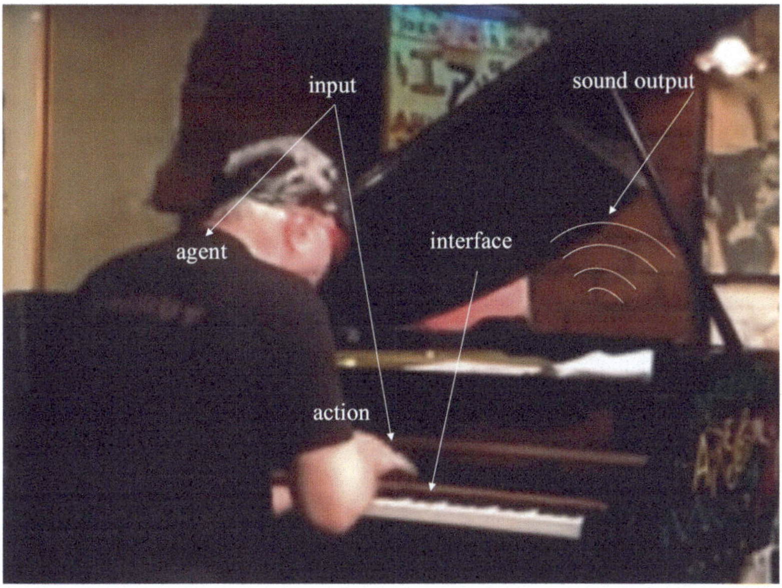

Fig. 7.3 The instrument's three components: input, interface, output. © Guerino Mazzola

Fig. 7.4 Görkem Şen's Yaybahar. Image reprinted with kind permission from Görkem Şen

3. New sound output
   - New sound generators (overtones, frequency modulation, physical modeling, samples, etc.)
   - New sound propagators (loudspeaker systems, spatialization, global networks) (Fig. 7.4)

Chapter 8
Musical Distribution Channels: New Networks

8.1 A Conceptual Understanding of the Evolution of Music Distribution in History

Long before the creation of recorded music, the consumption of music relied on a completely different set of parameters to determine its success. Consumers have grown accustomed to this reality, which was created by privilege to access, that is to say, most do not truly comprehend the evolution of the distribution of music throughout history before the advent of devices that replace musicians to create music. Unfortunately, due to the prevalent corruption of certain practices in the work of historians and music distributors, we have a small and regionally biased set of resources from which to pull historical information. We will aim to create a unified understanding of the earlier stages in the evolution of networking in music based on knowledge born from the combination of observations from regions of Europe, Africa, and Asia.

Language and music share a deep relationship in regards to the conceptual realization of how both spread and evolve. Both language and music exist because of the innate need for humans to engage in social practices. These social practices are completely reliant on a proximal existence. With that understanding, it becomes clear that like language, music originates in a proximal location. This is somewhat reflected by the current understanding of the largely different sounding music such as European Gregorian Chant, Javanese Gamelan, or West African Ju-Ju music. This proximal concept has stood the test of time, and even now, it is understood that musicians often first aim to spread their music to those that are close to them.

As time progresses, these proximal existences begin to expand, and a need arises for information to be recorded for a multitude of uses. The earliest music notation discovered is a Guqin music named Jieshi Youlan (碣石•幽兰) during the sixth or seventh century. The notation is named "Jianzi Pu," meaning the "written notation". Figure 8.1 shows one notation with explanation of each part. Musical notation shares an interchangeable relationship with its 'cousin', the written language. There

G. Mazzola et al., *The Future of Music*,
https://doi.org/10.1007/978-3-030-39709-8_8

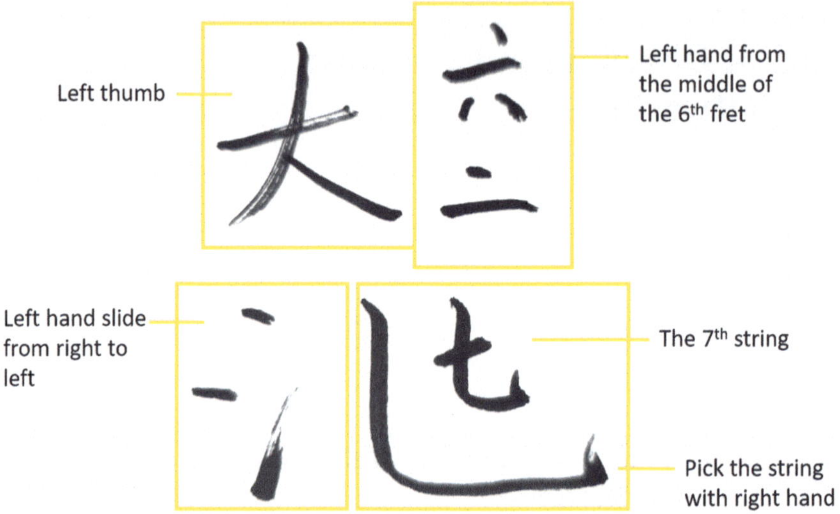

Fig. 8.1 Traditional notation for Guqin

are many benefits to the notational systems that were created through time, but it is not fair to ignore the harm it has caused in many aspects. There is now a way that music can be distributed throughout identities of proximity with efficiency, but as discussed in Chap. 14, there has yet to be a system that can completely capture all of the components and musical nuances to accurately recreate a music of the past without additional information. However, its importance is instrumental in retaining a fragment of the musical languages that have existed throughout history.

We have divided the evolutional points of music distribution development into proximity and written music. Realize nevertheless, that written music cannot exist without the requisite of proximal music performers. Until recently access to music was a privilege in many cultures, and to consume it one would have to search for those who can perform.

8.1.1 Pre-Internet Electronic Music Distribution Media: Phonographic Disc, Cassette, and Compact Disc

This situation is completely changed with the production of devices that remove the need for human interaction to create music. Technological advancements result in the creation of new instruments like the player piano, and further, devices that capture sound with the ability to reproduce those same sounds. The development of recording devices has also resulted in a need to create higher quality tools that can better store the information of the recorded sounds. The first popular storage

Fig. 8.2 The Phonographic Disc (Record, later also LP, Long Play) finds its origins in 1877 after Thomas Edison invents the phonograph and creates a disc shaped storage device to store information

Fig. 8.3 The tape cassette was introduced in 1962 by Philips, using the magnetic recording technology based on information introduced by Valdmar Poulsen (a Danish engineer) in 1898

devices belonged to the family of analog storage systems such as the phonograph record, see Fig. 8.2, and the tape cassette, see Fig. 8.3.

The popularity of the tape cassette ultimately outgrew its record counterpart due to its ability to re-record. This ability marks the beginning of the consumer to be able to obtain music illegally with ease. With access becoming even more readily available, the only thing left to improve was quality. This result was achieved by the creation and eventual expansions of the realm of digital technology and communications. The compact disc (CD) became the new popular music storage device as both quality and efficiency were key in its own development, see Fig. 8.4. The CD is tailored to match the finite hearing range of humans, which affords music

Fig. 8.4 The CD was developed by Sony and Philips in 1982. It uses a photodiode to read the differences in diffracted light, and converts that information to data

to be stored with high levels of quality, but eliminates the theoretical issues with capturing sounds that require an almost infinite amount of data to accurately record and reproduce in digital language.

This evolutionary point—LP, cassette, and CD—forms its foundation in removing the need to have access to trained musicians to perform. Without this need, access to music has become readily attainable to the point where the ease of access creates an overexposure that will be discussed later in this chapter.

8.2 Present Internet-Based Channels

The contemporary approaches of music distribution have been shaped dramatically by the recent internet revolution. The market of classical record labels has been gradually replaced by online music stores, such as *CD Baby*, which sells both physical CD copies and digital music. The market is later invaded by modern music streaming platforms such as *Spotify* and *Pandora*, available cross-platform on desktops, tablets, smartphones, and even in cars. And of course *YouTube*, one of the most beloved video services that combines music with other forms of media, is capable of enabling music creation interactions among a vast user base. As of 2018, more than 1.8 billion people registered on the site to check it daily to watch 5 billion videos. Every 60 s more than 300 h of video are uploaded to YouTube, which has a massive collection already of 1,300,000,000 videos.

Such evolution of music distribution has supported and enhanced common traits of modern lifestyle by introducing flexibility, and adaptability, compared to early methods like CDs and vinyls. It can be beneficial or harmful in terms of perspectives, unfolded in the following sections.

The first group rewarded by adopting the new distribution practice, especially with regard to the cost of distribution were the music producers. Think about how much of the burden they were carrying in the past: the manufacturing cost, and the enormous time wandering around the street advertising their music and so on. Musicians also felt this burden as they were often forced to depend on record labels, signing enforced contracts because they couldn't handle the cumbersome complexities in distributing their music. None of them are required nowadays thanks to modern technologies. With a bit of Googling, and a few clicks on your laptop, all of a sudden, you have the ability to greet audiences across the world. Not to mention that one can share the link of one's music with one's friends and followers on Facebook and Twitter with *zero cost*!

Nevertheless, things are more complex for audience. Bright sides and dark sides coexist in Internet-based music distribution. Among them, storage cost is one of the obvious benefits in the audience standpoint. We all know that physical CDs and vinyls consume storage space. Moving and storing, even maintaining these operations' condition was a challenge in the past. Lucky for us, dozens of years have witnessed physical space of CDs shrink into small digital files on our laptops, in the form of MP3 or WAV files. And now music is everywhere. Music and video streaming via cloud and distributed networks becomes the new king of music distribution. People can listen to their favorite bands so long as there is an Internet connection with *no* storage overhead.

What is the dark side of the story? There is no longer collective value to music. Your father's favorite vinyls, the CD you received at your high school prom, the special collections that couldn't be found in the current market. It is not just about the music they store, it is also about memories, emotions and all things connected for their uniqueness and the symbolism behind their physical form, whereas in streaming, replicability is unavoidable and music is just too easy to get, thus decreases its meaning in some way.

Besides the contradiction between optimized storage cost and decreased collectible values of music, as an audience, another interesting opposing argument is raised by the comparison between guaranteed ownership of music in the past and how we own music now. Many music endeavors will be lucky to be born in a world with enormous online databases. They cover countless genres, whereas buying physical copies of music can't provide this incredible amount of choice. However, some people find it to be a sad thing, for it seems impossible to own music in this streaming distribution. Think about the scenario when you appreciate one artist from Spotify while another in Apple Music. The only way to get them is to subscribe to both services, and forever be a subscriber. If you stop paying, you lose the music you love. If you love music from multiple platforms, then unfortunately, you have to pay for all of them. As an audience, we are always renting music online, we don't truly own any of our online music.

Let us reveal some insights on the impact of Internet-based distribution on the market. Economic value is at the heart of the operation of modern society. With the combination of music and media on YouTube, Internet-based distribution is not only helping economic market through selling or renting music, but also through promotional advertisements. Most music YouTubers work with large brands and insert ads in their music videos here and then. Spotify plays ads to non subscripts. The list just goes on. Also the transaction cycles of market are much more rapid than before. People used to spend years working on their music records, and wait for months to gather feedback from sales number and critical reviews. But now? The moment you put your new music on the Internet, you can get comments from real people. One cycle of preparing, selling, gathering feedback becomes much shorter, which in turn guarantees liveness of market and tolerates more trials and errors, thus leads to more success in the market.

If the above two are just improvements, the next two changes are more likely to become innovations and enhancements. Firstly, Internet-based distribution shapes the music industry by stimulating a new industry, namely, the online music library. It provides a platform for both professional and amateur musicians by renting their music/musical ideas to media creators whenever the uniqueness of music is not needed. This is so special in the sense that the purpose of such distribution is more like a distribution of resources, which later be used in someone else's work. That can't be done in the traditional way of music distribution since there is no way to track the usage of music in physical form. So the music creators simply banned their audience to adapt their music in any format, other than covers. Furthermore, this newborn industry serves both small-budget media creators and musicians especially orchestrators, whose needs were seldom satisfied in traditional music distribution. A similar use case is the libraries for shareable music elements, such as uploaded a cappella vocal tracks, EDM beats, sampled music, etc. Instead of serving media creators, this type of distribution targets at secondary musicians who makes music out of existing music. Or one can understand this in a way that the Internet-based distribution triggers collaborations between musicians far apart. With such resource pools, musicians don't need to create everything from scratch. This kind of resource redistribution is something that the old music distribution can't accomplish.

However, Internet-based distribution does its harm to the market as well. In the past, people could sell their CDs and Vinyls. The older they get, the more precious they become. Physical copies actually grow its value in time. These are the general trends for physical forms of creativity. Nevertheless, with the newborn Internet-based distribution, this type of value is killed before birth. But considering all bargains that the new distribution has brought to the table, this one flaw might still be forgivable.

8.3 Ubiquity and Omnipresence: Effects on Music Consumption Styles

The music's presence on the Internet now is everywhere (ubiquity) and always (omnipresence), like a god in religious environments or water from an environmental perspective. And the consumer can also access presently 36 million titles for example on Spotify. This unlimited availability is however not only given as a passive repertory, which the consumer may search like a traditional library. The presence is also an active penetration into personal data. For example, in collaboration with the US based *AncestryDNA*, which generates one's genome, playlists are now generated that fit one's DNA. The user profile also tracks the hour of the day and the weather to generate the user's potential preferences, such as "Ironic French-Language TV-shows" (for Spotify's movie repertory), which will be offered to the consumer on demand. The dating app *tinder* harmonizes its clients' music taste, and the *Cosmic* playlist generates a musical horoscope. Spotify has purchase the *Echo Nesto* in 2014, a US music intelligence platform developed at the MIT Media Lab. It has generated the largest playlist system ever, and adding to this the online platform *Tuningo*, Spotify can now construe playlists for all moods and activities of its users. The recent contract of the German music software company *Endel* with *Warner Brothers*, which we discuss in Sect. 19.1, is yet another development towards a machine-generated control of human moods in music.

This implies that the ubiquity and omnipresence of music also extends to its users' personal characteristics. The privacy is eliminated and replaced by a remarkable transparency of the human addressees. The elimination of privacy, and even of an individual's very definition of existence, is now being infiltrated on the level of one's musical identification.

This psychologically (moods) and physiologically (DNA) invasive service has generated a consumer behavior that increasingly eliminates willful decisions and replaces them by automatic preferences, which is very comfortable, well: comfortable like being a living dead zombie. Is this where they want us to move?

8.4 The Global Village of Music as Reshaped by Algorithms

What we recognize within these new directions is a "symbiosis of human and machine realities". More precisely: the increasing control of human behavior by algorithms, which are machine-driven devices using human behavior data to generate music for human 'outlets'. As discussed in Sect. 3.1.3, self-awareness, i.e., when music assists people to think about who they are, who they would like to be, and how they manage their own path, is a fundamental function of music which should not be controlled by and delegated to big brother machines.

Let us recall that music is a construction of an artificial reality, and most prominently of musical time (see Chap. 13). The global village should not castrate human creativity in favor of a socially flattened consensus on the nothingness of a numerical majority. Algorithms cannot think: they have no semiotic depths, they only process pattern matching operations.

So what can we propose in a musical environment against such a machine-business tyranny? Here are some proposals:

1. Replace machine driven creativity simulations by human instances.
2. Recreate a musical culture of living time creation in concerts, not in the Internet abstraction.
3. Invent music software the enables PC as a *Personal Creativity/Composition for everyone*, not only for experts.
4. Fight the tribal behavior that now dominates the music industry [134].

Chapter 9
Big Science in Music

9.1 Introduction

What is Big Science? It is the type of science which was introduced by the physicists via big particle accelerators together with its financial, sociological, and political implications. To give an example: the CERN (= Centre Européen pour la Recherche Nucléaire) with a total of roughly 10,000 employees embodies—among others—a Large Hadron Collider (LHC), the world's largest and most powerful particle accelerator in a ring of 27 km circumference; CERN's electricity consumption is comparable to the total volume of the city of Geneva, and the total annual financial volume is about one billion Euros [53]. Big Science is recognized by the following characteristics:

- Scientific language
- Models and theorems
- Experimental paradigms and operationalization
- Universal collaboration and communication
- Adequate laboratories and machines
- Political acceptance and corresponding resources

It is no coincidence that the universal communication on the internet was given the adequate streamline interface of world wide web in 1989 by CERN's Tim Berners-Lee on NeXT computers. In what follows, we want to make clear that music theory, too, needs and deserves the extension to a variant of Big Science.

G. Mazzola et al., *The Future of Music*,
https://doi.org/10.1007/978-3-030-39709-8_9

In order to actualize the historical link, let me cite Pierre Boulez, the founder and architect "malgré lui" of the IRCAM (= Institut Recherche Coordination Acoustique Musique) in Paris[1] [16]:

> Es ging mir bei meiner methodischen Untersuchung des musikalischen Universums und bei der Ableitung der vielfachen Konsequenzen, die aus einer Anzahl rationaler Ausgangspunkte zu ziehen sind, um den Versuch, ein zusammenhängendes System aufzubauen. Die Untersuchungsmethoden und die Erforschung eines solchen zusammenhängenden Systems halte ich für unerlässlich zur Grundlegung jeglicher schöpferischen Arbeit. Man halte mir nicht entgegen, eine derartige Auffassung führe zur Trockenheit, töte alle Phantasie und alle Inspiration – im Gegenteil: ich erblicke darin die stärkste Form der Erfindung, bei der die Imagination eine äusserst wichtige und bestimmende Rolle spielt. Wenn ästhetisches Denken unabhängig von der Wahl, der Entscheidung über die Technik, auftritt, kann es nur zum Bankrott führen: die Sprache wird, je nach Temperament, zu einer mehr oder weniger gewundenen Heuchelei oder zu einer banalen Gestikulation hersabgewürdigt...
>
> Unsere Imagination schärft die Intelligenz und unsere Intelligenz ist der Rückhalt der Imagination – ohne diese Wechselwirkung wird alle Forschung und Untersuchung zur Schmäre. Oft schon wurde gesagt: die Musik ist sowohl Wissenschaft wie Kunst. Wer könnte diese beiden Wesenheiten im selben Tiegel verschmelzen, wenn nicht die Imagination, diese 'Königin der Fähigkeiten'![2]

Evidently, Boulez's reflections are no plain statements but rather elements of a program which aims at realizing imagination such that arts and science are effectively reconciled. Let us state this implicit program in a way as to turn information technology into the central point of its feasibility:

Postulate 1 *Boulez's invocation of the "royal imagination" can only be envisaged by the virtual realization of the complex theoretical as well as practical system of music, its sounds and their relations by means of today's information technology.*

Recall that "virtual" means "being such in essence", in other words: Virtual realization means essential realization!

---

[1]There is an interesting anecdote I learned from the former Max-Planck director Valentino Braitenberg: Boulez was offered to direct the creation of a Max-Planck institute for music, but when he requested the possibility to work with orchestras and creative composers, not only doing traditional research with 'abstract' acoustical topics, they declined. The French president George Pompidou heard about this missed opportunity and gave Boulez the now successful IRCAM.

[2]My methodical investigation of the musical universe and my derivation of the multiple consequences to be drawn from a number of rational starting points was aimed at trying to build a coherent system. The methods of investigation and the exploration of such a coherent system are, in my opinion, indispensable for the foundation of any creative work. One should not hold against me that such an understanding leads to dryness, kills all imagination and all inspiration—on the contrary: I see in it the strongest form of invention, in which the imagination plays an extremely important and determining role. When aesthetic thought occurs regardless of the choice, the decision about the technique, it can only lead to bankruptcy: language, depending on one's temperament, is degraded to a more or less tortuous hypocrisy or to a banal gesticulation...

Our imagination sharpens the intelligence and our intelligence is the support of the imagination—without this interaction all research and investigation becomes a chimera. It has often been said that music is both science and art. Who could merge these two beings in the same crucible, if not the imagination, this 'queen of abilities'?

However, we are dramatically far from this objective, at least in German speaking Europe. For instance, in Switzerland, elementary particle research received roughly seventy times more financial support in 1994 than musicological research (multimedia projects included). And the presence of German speaking countries at the International Computer Music Conference '94 in Aarhus was remarkably weaker than anyone of the Latin, Scandinavian or Anglo-American country groups.

Nonetheless, music is a central issue in human life—though it affects another layer of reality than physics. We believe that the attempt to understand or to compose a mayor work of music is as important and difficult as the attempt to unify gravitation, electromagnetism, weak and strong forces.

And isn't it equally hyper-ambitious to unveil the God of physics through a world formula as to understand Beethoven's Fifth, for instance?! For sure, the ambitions are comparable, and hence, the tools should be comparable, too. So, where are we on the way to such an emancipation of Humanities from Natural Sciences?

9.2 Language, Models and Theorems

Let us get off the ground with a sketch of the present vectors towards a valid scientific language in music theory. In the concluding section of [16], Boulez answers the question why he so often refers to the analogy with the mathematical method:

> Weil die Mathematik die Wissenschaft ist, die zur Zeit die am weitesten entwickelte Methodologie besitzt, war mir daran gelegen, sie zum Vorbild zu nehmen, das uns helfen kann, unsere gegenwärtigen Schwachstellen zu beheben. Ich wollte gewissermassen die Fundamente für eine musikalische Methodologie legen.[3]

Since 1960, when this confession was expressed, Boulez was completely confirmed by the developments which are termed "Computing Musicology" or "Mathematical Music Theory", depending on the trans- resp. cisatlantic preferences. The tendency is readily recognized among the methodological spectrum which is presented on the occasion of the major conferences like the ESCOM (European Society of Cognitive Science in Music), or the ICMC (International Computer Music Conference).

To be clear, we agree with Boulez that music cannot degenerate or reduce to a section of mathematics: music is fundamentally rooted within physical, psychological and semiotic realities. But the formal description of musical instances by far transcends purely statistical tools for empirical data analysis. This extension emerges from the operationalization of large portions of the mathematical formalism by today's information technology.

A more in depth analysis of the increasing presence of mathematics in music theory during the last 150 years reveals a surprisingly parallel evolution of these

[3]Because mathematics is the science that currently possesses the most advanced methodology, it was important to me to take it as a model that could help us correct our present weaknesses. To a certain extent, I wanted to lay the foundations for a musical methodology.

sciences, an evolution that clarifies much of the more recent phenomena. Let us shortly sketch four major paradigms of both music theory and mathematics:

- **Global gestalts**: Musical as well as mathematical objects are often composed as patchworks of local structures. In music theory this was first made explicit by Eduard Hanslick [42] around 1854, an insight that music theorist Hugo Riemann called a "unity within the manifold". In mathematics, in the same year, mathematician Bernhard Riemann introduced the mathematical concept of a manifold [141], a patchwork of local cartesian charts.
- **Symmetries**: Symmetries were prominently investigated in 1924 by mathematician and music theorist Wolfgang Graeser in Johan Sebastian Bach's *Kunst der Fuge* [41]. Here is his summary:

  > Die Eigenschaft der Symmetrie spielt in der Musik eine so ungeheure Rolle, dass sie verdient, an erster Stelle betrachtet zu werden. Wir werden in der 'Kunst der Fuge' ihre fast uneingeschränkte Herrschaft besonders deutlich erkennen.[4]

- **The Yoneda philosophy**: In 1954, precisely 100 year after the rise of the global gestalt paradigm, the revolutionary idea of "understanding by changing perspective" was introduced in musicology and mathematics. It deals with a general methodology of understanding. In its common sense rendering, it states that in order to understand an object, you just have to walk round the object. This seemingly trivial insight was put into a rigorous mathematical shape by Nobuo Yoneda [159]. His famous Yoneda lemma states that a mathematical object can be classified by the system of all 'views' or 'perspectives' of the given object from all other objects of the same structure type (category). In music theory, the complete understanding of a composition (the typical object of this science) is a delicate process. This is due to the ontological complexity of a work. In fact, the score is only its first 'view', and it has to be completed by a variety of interpretations and performances in order to grow into a fully flagged identity. There is no unique optimal performance—this is a commonplace. The work's identity has to result from the interrelations of the (possibly infinite) variety of analytical and synthetical readings. Within this manifold, the composer's view is but the very first one—and not always the best. This insight backed the idea of Theodor Wiesengrund Adorno when he stated this also in 1954 [2]: *Die Idee der Interpretation gehört zur Musik selber und ist ihr nicht akzidentiell.*[5] This means that the entire system of interpretative perspectives of a given work is essential for the constitution of its being. In other words, we can suggest that Adorno's performance theory duplicates the synchronous lemma of Yoneda.
- **Gestures**: Gestures in music theory were introduced 2007 in [115] and initiated a new type of music theory, based on topological rather than algebraic arguments. Gestures are systems of continuous curves and resemble quite faithfully to what

[4]The quality of symmetry plays such an enormous role in music that it deserves to be considered first. In the 'Art of the Fugue' we will see its almost unrestricted rule particularly clearly.

[5]The idea of interpretation belongs to the music itself and is not accidental to it.

in physics are called strings. The musical gesture theory connects to homological algebra in mathematics, one of the most successful theories of the twentieth century.

Concerning models and theorems obtained by the research groups related to the authors, we refer to [118].

9.3 Experiments and the Operationalization of the Theory

In what follows, we want to illustrate the claim for Big Science in music theory by means of the analysis and performance workstation RUBATO®. Here are six important Big Science topics involved in its construction:

- Database management system research
- Hypermedia analysis and navigation
- High performance combinatorics
- Numerical differential equations
- Theoretical mathematics
- New methods in statistics

Let us discuss some of these points.

9.3.1 Database Management System Research

The representation and processing of musical score data is a delicate issue in the field of computing musicology. Problems do not only arise when capturing music with a given set of predefined signs, but also when translating these signs into a machine readable and processible format. This is due to the context dependent semantic flexions and indeterminations inherent to most natural human sign systems. By an in-depth study of the diversity of musical signs and relating this to structuralist semiology, we introduced the concept of "denotators" and "forms" to our database management system. Denotators are a kind of points in abstract spaces, called forms. This conceptual architecture is discussed in Sect. 10.1.1. It has evoked a pronounced interest by philosophers since it proposes a complete, programmable, and extensible conceptual environment.

9.3.2 *High Performance Combinatorics and New Methods in Statistics*

We want to put into evidence a central point for understanding the hitherto almost purely qualitative, deficient methods of traditional musical analysis: crashing because of desperately ample combinatorics arising from detailed listing and handling of elementary musical gestalts.

Motive cardinality in $\mathbb{Z}_{12}^2$	Number of isomorphism classes
1	1
2	5
3	26
4	216
5	2024
6	2780
7	41,720
8	634,573
9	9,059,071
10	119,032,295
11	143,038,358,371
12	157,430,569,051
13	1,592,645,620,686
14	14,873,235,105,552
15	128,762,751,824,308
16	1,037,532,923,086,353
17	7,809,413,514,931,644
18	55,089,365,597,956,206
19	365,290,003,947,963,446
20	2,282,919,558,918,081,919
...	...
72	2,230,741,522,540,743,033,415,296,821,609,381,912

Let us make a concrete example: Schumann's *Träumerei*, No. 7 of his *Kinderszenen* Op. 15. To illustrate this example, we use the motivic analysis component MeloRubette in our software RUBATO®. And here are the numbers. We choose motives with onset difference between first and last note at most a half measure:

- 1-motifs (single notes): 463
- 2-motifs: 3403
- 3-motifs: 9711
- 4-motifs: 12,631

i.e., a total of 25,745 proper motifs. The evaluation of the motivic context resides on the pairwise comparison and comprises 536,524,569 comparisons of motifs! The fundamental problem is that the collection of submotifs of a realistic composition is a combinatorial phenomenon of incredible complexity. Already for the subsets of 8 events within the *Träumerei*, we get a number of $4.92 \cdot 10^{16}$ candidates! And no scientifically valid procedure is known about how to select a reasonably small number of motives from such a quantity.

Let us give a more representative calculation: the number of isomorphism classes of event sets within the space of 12 pitch classes and 12 onset classes (integers modulo 12). The isomorphisms are defined by affine transformations of the ambient space, for technical details, we refer to [118, Ch. 11.3.8]. These class numbers have been calculated by Harald Fripertinger [36]. They are shown in the above table. The last row refers to the motives with half the total of $144 = 12^2$ positions. This number of size 10^{37} is comparable to the total number 10^{40} of elementary particles in the universe.

This result makes plausible that, to control motives and other subset types within a realistic composition of some thousand tone events, not only 463 as in the above composition, the combinatorial performance of software and computers has to be massively improved—and this asks for Big Science environments—and we have to develop intelligent statistical tools to select samples from the exorbitant collections of motifs and their classes. The latter procedure is an alternative since there is no chance for any computer to control combinatorics for large compositions, like a sonata, for example.

So this subject is a challenge for supercomputer combinatorics and statistics of large combinatorial ensembles. And we now understand why classical music theory was not in state of doing serious analysis: it lacked the concepts and the tools. Neither similarity concepts of gestalts nor a thorough contextual analysis were in reach of that tradition.

9.3.3 Laboratories

It is mandatory to extend this Big Science research in music to experiments with real music, compositions, new orchestrations, new musical instruments, and acoustic environments. One example of such a laboratory is the IRCAM in Paris, where Boulez, its first director, insisted on such experimental aspects. One should imagine a thorough analysis of Beethoven's *"Hammerklavier" Sonata Op. 106* (some 40,000 notes) and, according to Boulez's creative analysis, a recomposition of the sonata's musical substance (see also Chap. 12 for such an idea) for a large orchestra, possibly with new electronic instruments and even robotic musicians. Why would this be less interesting than the search for yet another elementary particle?

9.4 Political Acceptance

In physics, Big Science is politically accepted, the CERN and similar global institutes are a reality. One might ask for reasons behind this fact. Evidently, physics is not only an academic endeavor, but implies innumerable industrial and military applications. The atomic bombs on Hiroshima and Nagasaki are a tragic example thereof. The internet is a less tragic application.

Why shouldn't this also be valid for music? The answer might seem evident: no direct industrial and military applications. What about mathematics? It is the backbone of all hard sciences and does not rely on laboratories with experimental devices, except for some rather harmless computers for numerical calculations. Music theory in its complexity can be compared with mathematics, but it does not directly apply to hard sciences. This is however not completely true. We should recall that many ideas and methods in mathematics and physics were modeled in/by music, see Chap. 4. And we should keep in mind that sociology and psychology of humans are heavily influenced by our musical culture. The profit from music is not disjoint from our culture. But it is an interesting question why Big Science in music should not be supported like Big Science in physics. After all, industry and military power should not be an ultimate priority of human evolution.

Part III
Mathematical Concepts

Chapter 10
Mathematical Music Theory

Mathematical Music Theory, MaMuTh in short, is a term that was introduced by Mazzola in 1981 to specify a branch of music theory which would use mathematical concepts and theorems as opposed to traditional[1] music theory, where concepts and methods were dominated by a mix of philosophical, philological, and psychological or cognitive perspectives.

The necessity of such a specification was driven by two forces, (1) the introduction of computer technology (such as synthesizers or notation software) to music and (2) the failure of traditional approaches to solve fundamental problems in music theory and performance. For example the failure of harmony (theory), leading to twelve-tone (dodecaphonic) methods (see also Chap. 11), or counterpoint, leading to the "emancipation of dissonances". Historically viewed, European music theory started being a mathematical science, a fact that is also present in the Medieval canon of Liberal Arts, where mathematics and music were grouped in the *quadrivium* of arithmetics, geometry, astronomy, and music. They were separated from the *trivium* of grammar, logic, and rhetoric. Therefore MaMuTh is a revival of the original approach, but now using contemporary mathematics, not only elementary arithmetics.

10.1 The MaMuTh Components

MaMuTh is composed of *three components*, which are shared with all hard sciences. *The first is a language* that aims at describing the field's objects and relations in a precise and adequate way. Such a language was absent in traditional music theory. For example, the concept of "tonality" was fuzzy, ranging from scales to triadic harmony or single "tonic pitch", etc. Typically, many concepts were

[1]Such as tonal, post-tonal, and Schenkerian theory.

G. Mazzola et al., *The Future of Music*,
https://doi.org/10.1007/978-3-030-39709-8_10

homonymous, such as "cadence" or "inversion". Or the concept of "tempo", which can signify the relation between symbolic score notation of duration and physical duration in performance, as specified by the Maelzel Metronome number, but it can also mean the "density of sound events per minute". Such fuzzy conceptualizations are not simply common language, they testify a theoretical deficiency. For example, the notation of chords by altered pitch instead of pitch creates a blurred discourse, especially useful for theoretical approaches such as Schenker theory, where alterations help "prove" arbitrary tonal perspectives.[2]

The second component is the presentation of models for specific musical processes, such as harmony, melody, rhythm, counterpoint and voice leading, or performance. Such models are built upon the precise language, but also include theorems about the formal structure of such models. For example, a model of tonal modulation should specify the pivotal chords in the form of a theorem that describes such pivots, see also Chap. 12.

The third component is the experimental test of theoretical models, also essential with hard sciences. It relates theoretical thoughts with the domain of musical reality they claim to clarify. This component was radically neglected when dodecaphonic methods were introduced for the following reasons. The experimental material of harmonically complex composition by the end of the nineteenth century was left altogether because the theoretical models didn't work. Imagine a physical science where you would not reconsider experimental material because your theory doesn't work! You would have to shut down the entire CERN[3] enterprise. Realizing compositions means—besides the creative challenge—to make experiments with one's thoughts, emotions, and gestures. And music theory has to interact with any such experiment: evaluating given experiments and proposing new ones.

10.1.1 The Language in MaMuTh

The development of a scientific language for music theory in MaMuTh was also driven by the necessity to perform computer-aided computations for musical performance models [111, Part IV]. The result of this effort is the language of denotators and forms. Denotators represent musical objects, while forms are the spaces where denotators live. This conceptual framework uses cutting edge mathematics, especially topos theory, which is a marriage of geometry and logic, and has been used as a foundation of computer science and physics. At the bottom of the space architecture we have so-called *simple* forms. Simple forms are mathematical spaces, such as the real line $\mathbb{R}$ for onset or duration, or the pitch

[2]Recall a result by Mason and Mazzola [107, Ch. I.4]: Any two seven-element scales can be alterations of each other by at most three ♯ and two ♭. The usage of too many alterations does not prove anything.

[3]CERN is the European Center for Nuclear Research near Geneva.

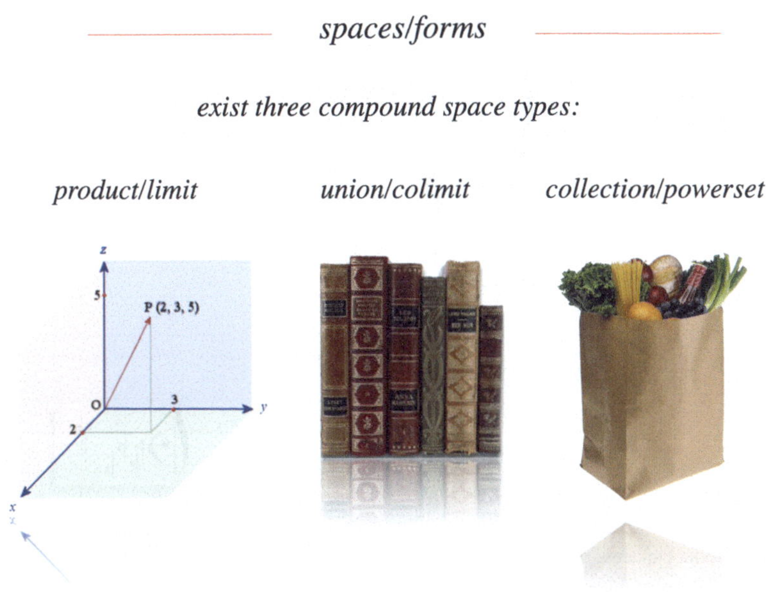

Fig. 10.1 Composed spaces can be products, unions, or collections of already given spaces. © Guerino Mazzola

class space $\mathbb{Z}_{12}$. But in our theory, concepts must all have an explicit name, this is an important difference to mathematics, where names are irrelevant. For example, a simple onset form would be notated by `Onset:.Simple(`$\mathbb{R}$`)`, and the duration form would be written as `Duration:.Simple(`$\mathbb{R}$`)`. One recognized the difference to mathematics: Same mathematical coordinates ($\mathbb{R}$), but different names in the musical setup. This approach uses three types of composed spaces from already given spaces, which are: *products* (also called limits), *unions* (also called colimits), and *collections* (also called powersets), see also Fig. 10.1.

First, products describe spaces by a number of coordinate spaces, and points in product spaces must be described by a sequence of coordinate values. For example in a standard score, a denotator is a note in the product space whose coordinates are onset, pitch, loudness, and duration. Such a product form would denoted by `Note:.Limit(Onset,Pitch,Loudness,Duration)`. A simpler note form is shown in Fig. 10.2. You can see there that denotators also have specific names, together with their form and coordinates. The denotator in Fig. 10.2 would be written as `myNote:@Note(myOnset,myPitch)`.

Second, unions are spaces that also have a number of coordinate spaces, but now the idea is akin to a library. The coordinate spaces are like books, and a denotator of a union is a denotator in one of the given 'coordinate books'. For example, in an orchestra, the instrumental parts are the books, and a note is a denotator in one of the given instrumental 'books'. See Fig. 10.3 for an example. Here, the form is named Orchestra, denoted by `Orchestra:.Colimit(Instrument1,`

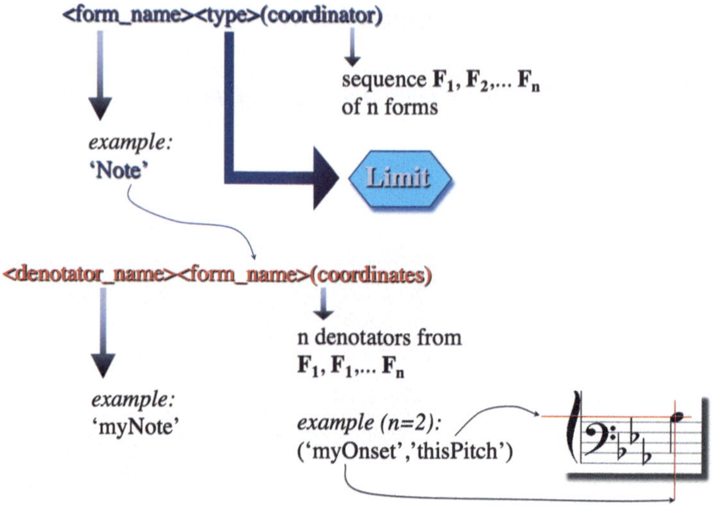

Fig. 10.2 Product spaces. © Guerino Mazzola

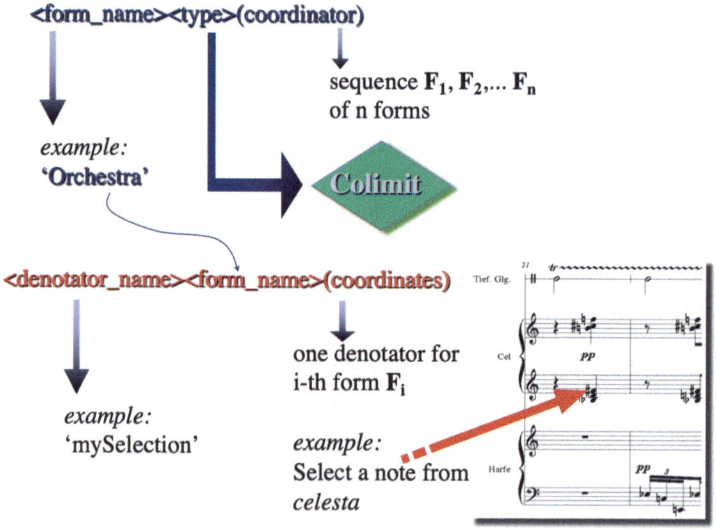

Fig. 10.3 Union spaces. © Guerino Mazzola

`Instrument2,...)`. It also refers to a collection of already defined forms (here Instrument1, Instrument2,. . .). But a denotator in such a union is defined by choosing one of the 'books' (Instrumenti = celesta, in our figure), and then one denotator in that book, here a note in the celesta part of the orchestra score.

Third, a collection space refers to one given coordinate space, and a denotator here is just a set of denotators in one single coordinate space. For example, a chord

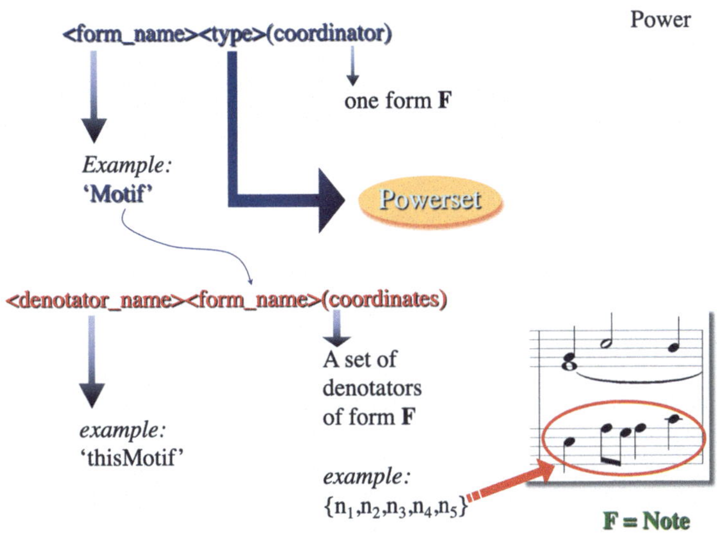

Fig. 10.4 Collection spaces. © Guerino Mazzola

is a collection of notes, which are denotators in the above product space of notes in a score. In our example in Fig. 10.4, the form is `Motif:.Powerset(Note)`, and denotator `thisMotif:@Motif`$\{n_1, n_2, n_3, n_4, n_5\}$ is a set of five notes.

This approach also includes spaces of musical gestures, which are not standard concepts of theoretical language in music. It has been implemented in the software RUBATO® for composition, music theory, performance, and gestures, see [118, Part X] and [113, Part XVII]. MaMuTh language has also been applied to a number of theoretical approaches, which are usually published in the *Journal of Mathematics & Music* and in Springer's book series *Computational Music Science*.

10.1.2 Models and Theorems

Models for composition, music theory, and performance have been established in MaMuTh and corresponding software. One of these models for counterpoint is described in Chap. 15. Another model for tonal modulation is applied in Chap. 12. Interesting models for harmonic analysis using Finite Fourier Theory have been developed by David Lewin, Ian Quinn, and Emmanuel Amiot, including corresponding software, see [9]. Computer-aided performance theory has been developed (among others) in Stockholm by Johan Sundberg, Anders Friberg, and Antonio Camurri [111, Ch. 13.5], and in Zürich by Oliver Zahorka and Mazzola [118, Parts VIII, IX, X]. The MaMuTh of performance is remarkable in that the philosophical terminology of Adorno and Walter Benjamin, specifying performance as the "infinite precision" in what they call "micrological procedure", could be made precise by use of classical methods of differential geometry, such as Lie operators. Here, MaMuTh was effective in translating even highly sophisticated philosophical concepts into an adequate mathematical conceptualization.

10.1.3 Experiments

A number of experiments in MaMuTh have been accomplished in different fields, such as harmony, counterpoint, or performance theory. The probably first experiment was the application of number theory to the classification of musical intervals by Leonhard Euler, where he could simulate the consonance quality of intervals in just tuning by his "gradus suavitatis" formula [121, Ch. 5.1.6]. The application of a model of tonal modulation to the modulatory landscape of the first movement of Beethovens *"Hammerklavier" Sonata op. 106* was another successful experimental test. The application of statistics in performance theory to famous performances of some classical compositions by Bach, Schumann, Webern, and Chopin by Jan Beran and Mazzola should be mentioned, see [114, Part XI]. Statistical methods together with contrapuntal models in MaMuTh have been applied to understand Alexander Scriabin's mystical chord in his *sonata No. 5* [6].

10.2 The Creative Power of MaMuTh

We should be aware of the creative power of mathematics in general and MaMuTh in particular. Mathematics has the most developed methodology of conceptual architectures, its development in the twentieth century has shown an explosion of new theories that are not simple calculation devices, but drive towards a mathematics of concepts. The precision and abstraction of mathematics and its collaborative globally accepted communication standard have given it the potent drive for physics, computer science, and MaMuTh to shape these sciences' future. Boulez in his writings [16] acknowledges this unique position as a model for music theory, see also Chap. 9.

Chapter 11
Serialism: Failure of New Concepts Without Musical Impact

In this chapter we want to discuss the compositional approach of serialism, and more concretely its realization with Pierre Boulez's prototypical composition *Structures pour deux pianos, livre I* [15] composed in 1952. Serialism was introduced 1947 by French composer René Leibowitz and taken over 1949 by the French composer Olivier Messiaen. It generalizes Arnold Schoenberg's dodecaphonic method of composition introduced around 1921.

Our discussion identifies some important drawbacks of serialism which are essentially due to the mathematical but musically insensitive nature on the one hand, and, on the other, to the absence of syntagmatic, harmonic, rhythmic, and melodic categories in its construction.

11.1 Principles of Serialism

Schoenberg's dodecaphonic method was a reaction to the increasingly complex and theoretically disoriented development of musical composition in the first decade of the twentieth century. The very concept of tonality was questioned, harmony was no longer a secure reference for the understanding and construction of musical meaning (see [118, Ch. 13.4.2.1]). Schoenberg decided to refrain from all traditional references and to propose a radically different compositional method, dodecaphonism. This approach does not specify harmonic, rhythmic or melodic/thematic structures, but proposes to build composition from a 12-tone row S. Such a row is an ordered sequence of 12 pitches, $S = (S_0, S_1, \dots S_{11})$, which represent all 12 pitch classes, each once. Together with this row, the method allows the composer to use the total of 48 transformed rows $T(S)$, where T is any combination of the 12 transpositions, inversions, retrogrades, and retrograde inversions. These 48 variants of the row S can be played in any patchwork, and also, every such row can be played with parts of the sequence at the same onset.

G. Mazzola et al., *The Future of Music*,
https://doi.org/10.1007/978-3-030-39709-8_11

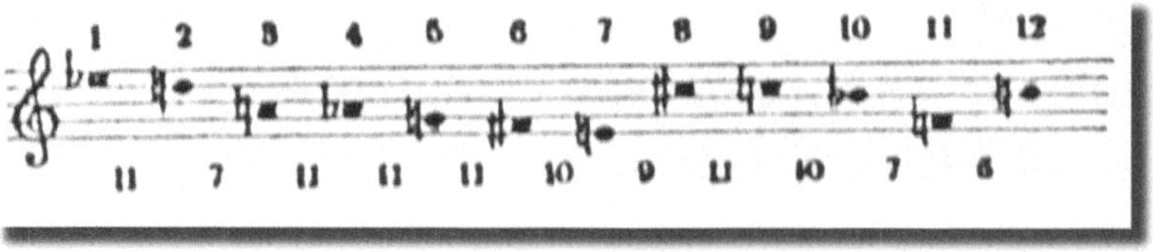

Fig. 11.1 The pitch series S_{pitch} in Boulez's serial composition. György Ligeti "Pierre Boulez" aus "Die Reihe: Heft 4 - Junge Komponisten" © by Universal Edition A.G., Wien/UE26104

For example, the row S may be played with S_0, S_1, S_2 at the same onset, and then S_3, S_4, S_5, S_6 one after the other, and then again S_7, S_8, S_9, S_{10} together, and terminating with the S_{11}.

The chords which would result from a simultaneous appearance of several notes from a number of transformed variants of the composition's row are in no way controlled by this method, also, the durations of the notes or their loudness are not controlled.

Schoenberg's idea was to give each of the 12 pitch classes the same relevance/context, there would no longer be a distinguished role of a tonic or dominant note. This is, however, not guaranteed, as may be shown by some special rows. Take the row $S = (C, D, E, F, G, A, B, C\sharp, D\sharp, F\sharp, G\sharp, A\sharp)$. The first seven pitches define a thoroughly C-major context, whereas the last five pitches define a pentatonic environment. Or take $S = (C, E, G, B\flat, D, F, A\flat, B, E\flat, G\flat, A, D\flat)$, which is a sequence of major and minor thirds and even starts with the C-major triad. Or one could build a row which starts with the Raga music's mela[1] "mayamalavagaula" $\{0, 3, 4, 7, 8, 9, 1, 2, 5, 6, 10, 11\}$ as initial portion. This defines a contextuality of pitches which is far from "neutral" as claimed by Schoenberg.

11.2 Boulez's Construction and Ligeti's Critique

The serial approach was initiated by Leibowitz and Messiaen. It consists of an extension of Schoenberg's idea to other musical parameters. A serial composition would not only start from a row of pitches, but also one of 12 durations, loudnesses, and attacks (Messiaen was composing for pianos, so the attacks were driven by this instrument's performance). Boulez took the row S_{pitch} from Messian's 1949 composition *Modes et valeurs d'intensité*, see Fig. 11.1.

[1] See also Chap. 15 for this Raga context in counterpoint.

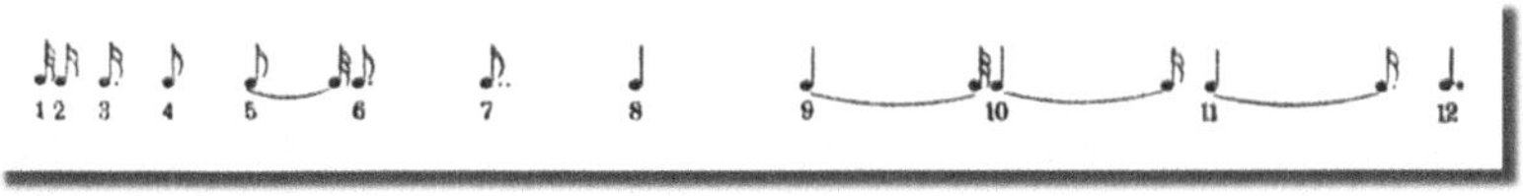

Fig. 11.2 The duration series $S_{duration}$ in Boulez's serial composition. György Ligeti "Pierre Boulez" aus "Die Reihe: Heft 4 - Junge Komponisten" © by Universal Edition A.G., Wien/UE26104

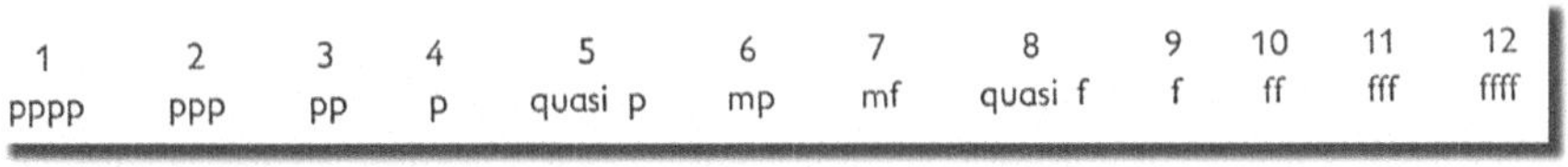

Fig. 11.3 The loudness series $S_{loudness}$ in Boulez's serial composition. György Ligeti "Pierre Boulez" aus "Die Reihe: Heft 4 - Junge Komponisten" © by Universal Edition A.G., Wien/UE26104

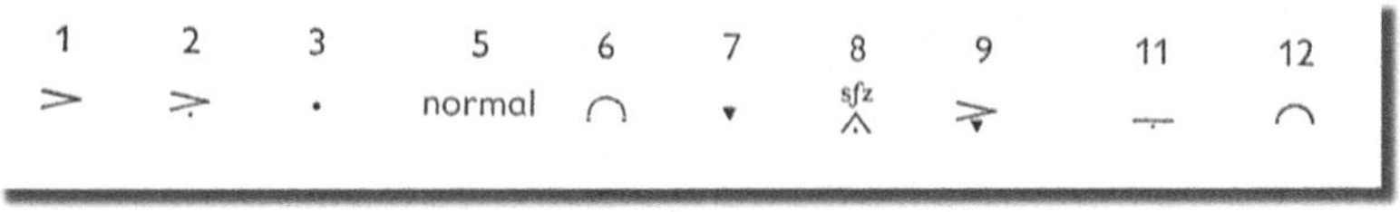

Fig. 11.4 The piano attack series S_{attack} in Boulez's serial composition. György Ligeti "Pierre Boulez" aus "Die Reihe: Heft 4 - Junge Komponisten" © by Universal Edition A.G., Wien/UE26104

For durations, Boulez took the series $S_{duration}$, as shown in Fig. 11.2. It is a completely arithmetic series with regularly increasing durations from 1/32 to 12/32. This implies in particular that every row would have a total duration of $1+2+\ldots 12 = 78$ 1/32 notes, since Boulez would not use rows with sequences of more than one note at the same onset.

The third series $S_{loudness}$ of loudness values is equally arithmetic, as shown in Fig. 11.3.

The fourth series S_{attack} of piano attacks is shown in Fig. 11.4.

Boulez would construct a number of variants of these rows, so-called "threads", for two pianos, and following a quite sophisticated mathematical construction. Figure 11.5 shows the stacks of threads in parts A and B of the composition. Figure 11.6 shows the first page of Boulez's score.

In a detailed analysis of Boulez's composition [102], composer György Ligeti criticized Boulez's procedure as being thoroughly "number fetishist". The problem for Ligeti was that he could not understand how Boulez would apply transformations for the rows of duration, loudness, and attack. In fact, the transformations of pitch, such as inversion or transposition, don't make sense on the three other parameters. For example, transposition of attacks is not conceived. But it can be shown (see our detailed discussion in [119, Ch. 25]) that Boulez came up with a very modern

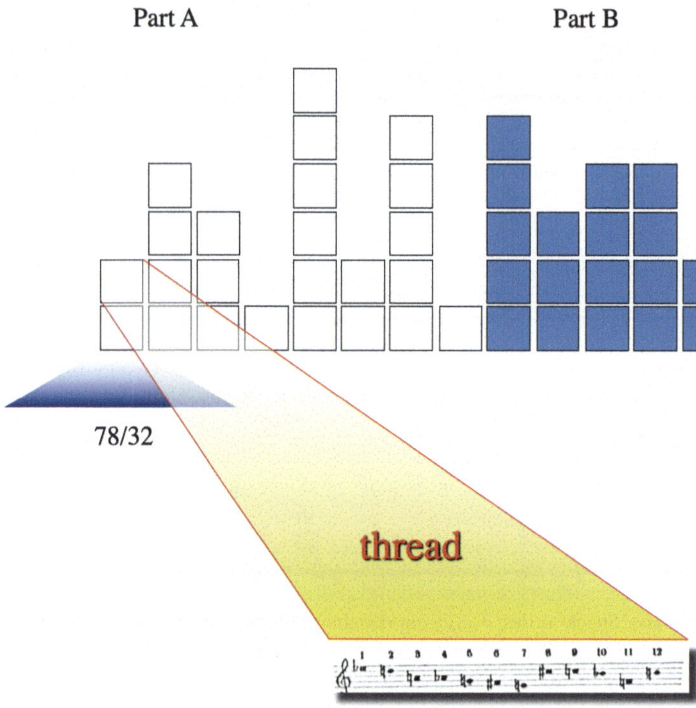

Fig. 11.5 The display of the rows, each block being one row, and every row having the same duration 78/32. © Guerino Mazzola

mathematical idea to solve this transformational problem. To this end, we represent a row as a map $S : \{0, 1, 2, \ldots 11\} \rightarrow P$, which means to define a sequence $S(0), S(1), S(2) \ldots S(11)$ of 12 different elements $S(0), \ldots S(11)$ in a parameter space P, see Fig. 11.7.

For pitch classes, this would be the clock $\mathbb{Z}_{12}$ of 12 h, for duration, it is the set $1/32, 2/32, \ldots 12/32$ of 12 durations, and so on for loudness and attack parameter sets. Boulez's ingenious idea was to understand that transformations for pitch classes can also be simulated by not transforming on $\mathbb{Z}_{12}$, but by prepending a transformation on the set $\{0, 1, 2, \ldots 11\}$ of arguments instead. For example, retrograde is just taking the reversed order, i.e., prepending the permutation $0 \mapsto 11, 1 \mapsto 10, \ldots 11 \mapsto 0$ to S. With this trick, Boulez could apply transformations also on the other parameters by simply prepending permutations on the argument

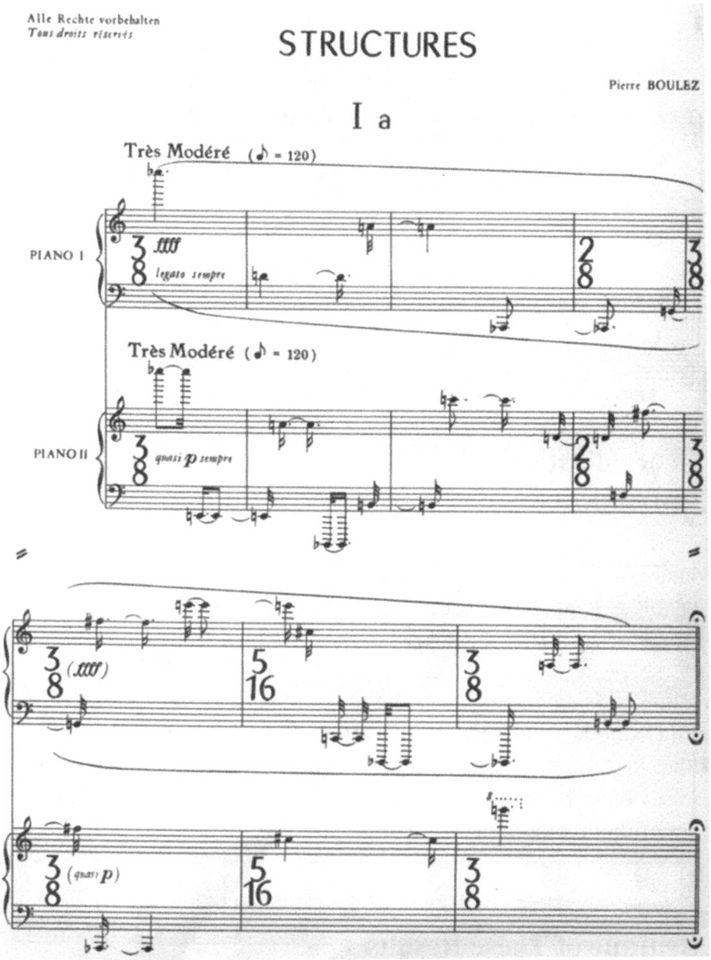

Fig. 11.6 Pierre Boulez "Structures|Premier Livre|für 2 Klaviere" (1. page of the score) © 1955 by Universal Edition (London) Ltd., London/UE12267

set. This idea to view collections in parameter sets as being maps from "address" sets of arguments has been a major progress in modern mathematics and has been applied with great success in mathematical music theory, see [118] for a detailed discourse. If we view Boulez's procedure using such addressed methods, his architecture becomes completely natural.

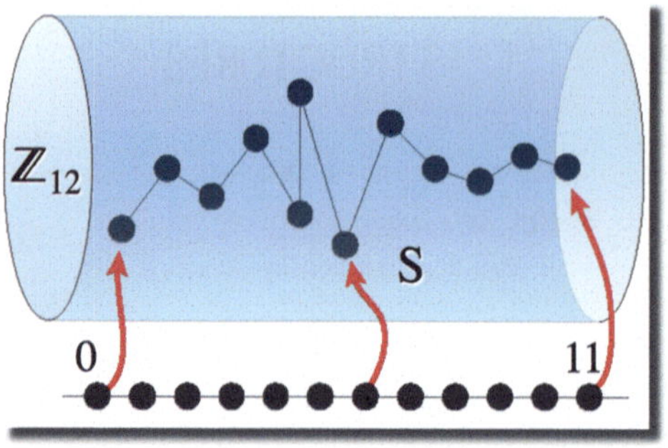

Fig. 11.7 A row is seen as a map from a set of 12 arguments, 0, 1, 2, ... 11, to a parameter set. © Guerino Mazzola

11.3 Generalization of Boulez's Construction to 12 Instruments

Ligeti's analysis and our mathematical interpretation of Boulez's compositional method enable a general setup of that method, which is no longer related to the special choice of rows and the two pianos. In [125, Ch. 18], we have described a RUBATO® software implementation of this generalization to 12 instruments. This generalization was acclaimed by IRCAM's audience when presented in a talk by Mazzola in December 2007 dedicated to Boulez.

After that first implementation, more flexible RUBATO® implementations with variable rows have been realized, which means that we now dispose of several realizations of that method for 12 instruments.

11.4 Critique of These Results

It is a historical fact that to this date serial compositions have never reached a broader audience (even an audience of music lovers who would appreciate other twentieth century composers, such as Bartók, Prokofiev, etc.). In what follows we want to discuss objective arguments backing this fact.

We should also add that none of the existing versions of this serial creation have ever reached any positive audience among our students and professionals of composition. Also, IRCAM's audience preferred the 12 instrument version to Boulez's original version, but this does not mean that the material as such was appreciated independently of Boulez being IRCAM's director.

11.4.1 Mathematical Abstraction

The very first observation regarding the idea of dodecaphonism and serialism is that it is a thoroughly mathematical one, and as such even a "baby theory": You simply set up a permutation of 12 musical parameters in a temporal sequence, the row, and you then apply 48 transformations to this row. This procedure has nothing musical apart from the permutational selection of musical parameters. This selection has no musical rationale. It is an application of a purely mathematical triviality to musical parameters without any consideration of their musical nature whatsoever. This poiesis is detached from music by its very definition. And, as observed above, its claim of putting all the parameter values into a "democratic" equal position is not achieved.

It is known that serial composers (Boulez for example in his book [16]) tried to introduce more detailed internal structures within given rows, such as inner symmetries or correspondences between parts of such rows. This might be seen as an attempt to enrich the naive mathematical row preset.

We may summarize this observation in saying that *the poietical mathematical technique is not as such a guarantee for a working musical structure. The formalism must always offer a musical realization.*

11.4.2 Absence of Ordered Syntagm

From the structural point of view, the serial approach does not imply any syntagmatic order, which would be present in traditional approaches, for example with the cadential syntax $I - IV - V - I$, or $2 - 5 - 1$ in jazz, which is not only the set of degrees, but also their fixed order. Also in a larger form perspective, such as the sonata allegro, the rhetorical syntax of exposition, development, recapitulation, and coda, together with the detailed sequence of tonal units, such as cadences and modulations, and the development of melodic/motivic themes, this is all absent in the serial approach. You may just throw those transformed rows into a random patchwork on the score. Meaningful syntactical structures are essential in semiotic systems. The syntactical construction of contents is not proposed in serialism.

11.4.3 Failure of Communication

Serialism also fails on an elementary level of communication. To begin with, the row is virtually never recognized from the auditory experience. Especially if we hear several rows from the beginning, as seen in Boulez's composition. Moreover, there is no compositional tool in the serial approach to make evident the connection of different transformations of the row. If we present first the original row and then a transformation, how could the listener recognize the relationship (the transformation) of these two instances? Beethoven in fact was efficient in showing, for example, his inversions *musically* in his compositions, see [118, Ch. 28.2.5]. The syntagmatically chaotic display of the row patchwork makes the recognition (even if one wants to analyze the graphical score display) of row relationships very difficult if not impossible.

11.4.4 No Harmony, Rhythm, or Melody

We are not referring to any traditional harmonic, rhythmic or melodic paradigms here. The point is that serialism has simply no relation to harmonic, rhythmic or melodic paradigms. Whereas Scriabin realizes new and different harmonic and contrapuntal structures, serialism has no such sensitivity, it is 'colorblind' for such aspects. The simultaneous appearance of pitches create chords without any meaningful content in the serialist perspective. Idem for rhythm and melody. The latter is formally represented by the row and its transformations, but the row as such has no melodic character whatsoever. Intervals can be consonant, dissonant, they simply don't enter into the serialist discourse.

11.4.5 The Sociological Role of Serialism

In view of the preceding discussion, serialism (and to some degree also its mother method of dodecaphonism) lacks communication of its trivial mathematical machinery, it is oblivious of any type of harmonic, rhythmical or melodic structure, and it has no syntactical order.

This method has *therefore* not been capable of establishing a successful position in the development of twentieth century composition. It has influenced many composers, but mainly by inhibiting their musical creativity, the approach resembles more a musical catastrophe than anything else. In view of this fact, it is mysterious why in academic music education serialism is still a dominant topic. It is remarkable that the introduction of jazz has become significantly more and positively influential for twentieth century music than the serial experiment, but it is rarely discussed in academic education.

Chapter 12
Mazzola's Sonata Construction: A Technical Approach and Its Limits

12.1 Boulez's Creative Analysis

Mazzola's sonata was composed in the spirit of Pierre Boulez's "creative analysis" [14]. Boulez's idea was to analyze a composition, but not for its own sake, which Boulez disqualified as a sterile exercise. He suggested using the analysis to compose a new piece that would be a variation of the given material. Boulez thought neither of deduction nor induction, but a lateral movement into a neighborhood of the given analyzed material, a movement which would pertain to what Boulez called an *open work*. In our case this means that Beethoven's *"Hammerklavier" sonata Op. 106*, Allegro movement, was analyzed, and this analysis was used to create a new composition, Mazzola's sonata Allegro movement of op. 3, published and recorded under the title *L'essence du bleu* [110] by changing some analytical parameters.

12.2 Applying Creative Analysis to Beethoven's *Hammerklavier Sonata Op. 106*, Allegro Movement

In our analysis of Beethoven's *"Hammerklavier" Sonata, Op. 106*, we exhibited the central role for harmony and motivic work of the diminished seventh chord $C\sharp^{\circ 7} = \{c\sharp, e, g, b\flat\}$, see Fig. 12.1. Our analysis therefore has a harmonic and a motivic aspect.

12.2.1 Modulation Theory

The harmonic aspect applies tonal modulation theory as described in [108], see [120, Ch. 23] for a short presentation of the theory. The detailed discussion

G. Mazzola et al., *The Future of Music*,
https://doi.org/10.1007/978-3-030-39709-8_12

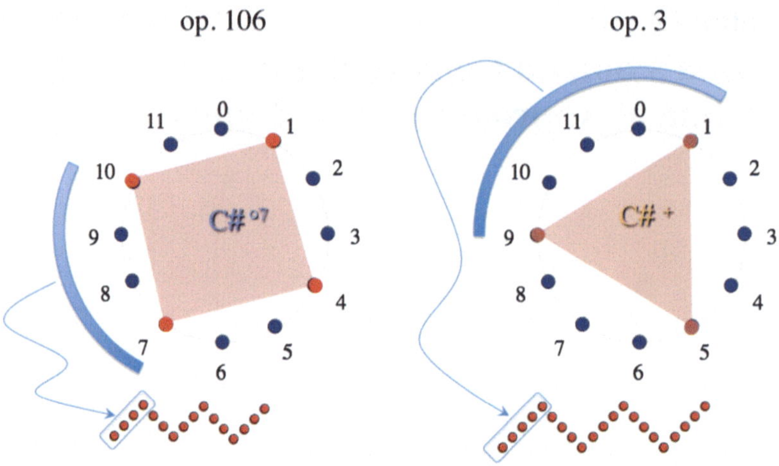

Fig. 12.1 The motivic zigzags in Beethoven's *op. 106* and Mazzola's *op. 3*. © Guerino Mazzola

of Beethoven's modulations in Op. 106, Allegro movement, is exposed in [118, Ch. 28.2]. Our modulation theory is based on symmetries, inversions or transpositions, that connect given tonalities. In Beethoven's case, it turns out that modulations are related to the inner symmetries of the chord $C\sharp^{\circ 7}$. These symmetries divide the set of tonalities into two orbits, the "world" of the eight diatonic scales of $B\flat$, $D\flat$, E, G, A, C, $E\flat$, $G\flat$, and an "antiworld" of four diatonic scales of D, F, $A\flat$, B. These two orbits show precisely what music theorists Jürgen Uhde and Erwin Ratz had called the "world" around $B\flat$ major and the "antiworld" around B minor. These symmetries also explain all modulations in detail.

In our study of Mazzola's sonata construction, we used this analysis to create a new modulatory architecture.

12.2.2 *The Generic Motive*

Apart from determining the sonata's harmonic modulations, the chord $C\sharp^{\circ 7}$ also determines Beethoven's motivic work. As the chord $C\sharp^{\circ 7}$ is reproduced after transposition by three semitones[1] T^3 of its pitch classes, it is reasonable to consider an elementary motivic "zigzag" of period 3. This is shown in Fig. 12.1 to the left, where a chromatic ascending and descending melodic movement is visible. In Op. 106, this zigzag is a dominating motivic structure [107, Ch. IV.3].

[1] A transposition by n semitones is denoted by T^n.

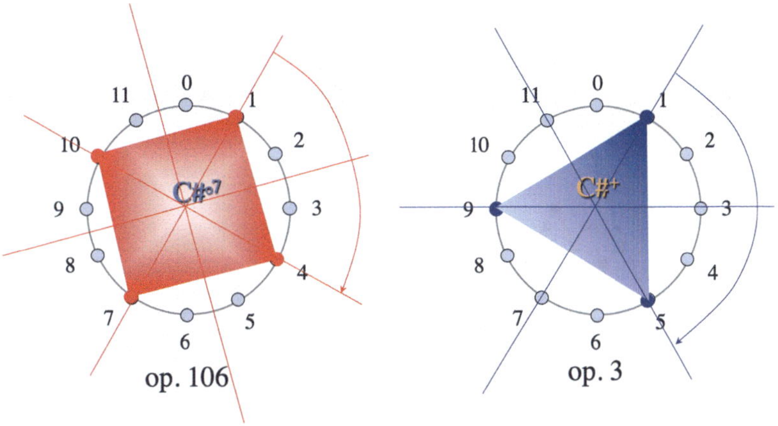

Fig. 12.2 The symmetries of the two fundamental chords, $C\sharp^{\circ 7}$ for Op. 106, and $C\sharp^{+}$ for Op. 3. © Guerino Mazzola

12.3 Transfer to a New Sonata Allegro Construction: Mazzola's *Op. 3*

In the creative part of our analysis, we replaced the minor third transposition T^3 by a major third transposition T^4 since $12 = 3 \cdot 4$ is the decomposition of 12 into different prime number powers. The corresponding chord is the augmented triad[2] $C\sharp^{+} = \{1, 5, 9\}$ that is reproduced after a transposition T^4 of pitch classes; see Fig. 12.2 for the symmetries of the two basic chords.

12.3.1 Modulation

To shape a modulatory architecture, we considered the symmetries of $C\sharp^{+}$. They divide the set of 12 diatonic scales—for major (ionian) or minor (aeolian) tonalities—into three "worlds": $\{D, C, B\flat, A\flat, G\flat, E\}, \{A, F, D\flat\}, \{G, E\flat, b\}$. The modulations of our sonata movement only work for symmetries that move tonalities within the three worlds.

[2]For music theorists: The sharpened/flattened note names do not mean that we think in alterations, but simply denote the position of the note on the set of 12 pitch classes within an octave.

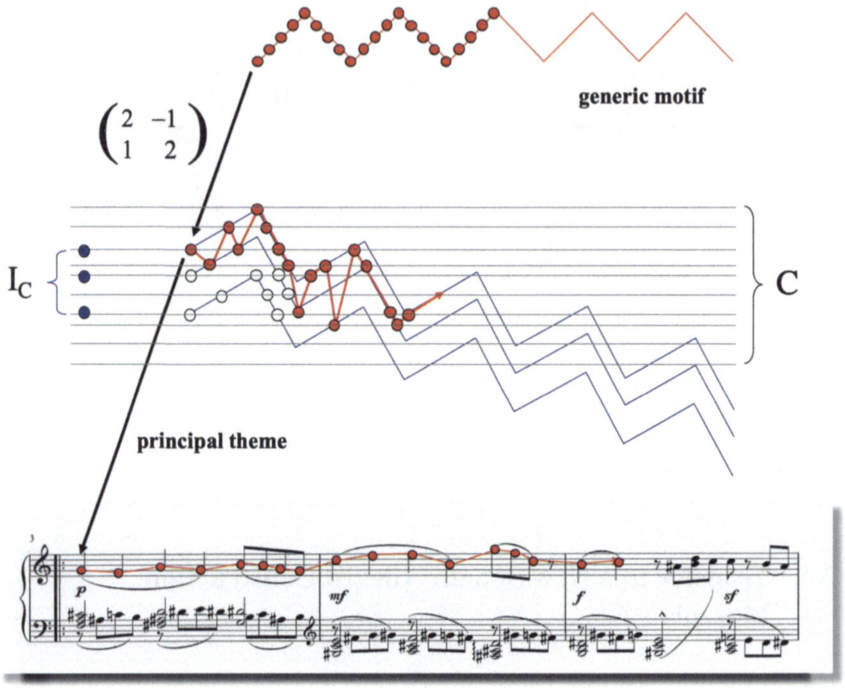

Fig. 12.3 The construction of the sonata's main melody from the generic motive. © Guerino Mazzola

12.3.2 The Generic Motive and the Main Melody

The corresponding motivic zigzag of period 4 is shown below the chord $C\sharp^+$ to the right in Fig. 12.1. The main melody of our sonata is also constructed from the generic zigzag and by no means from romantic intuition. See Fig. 12.3 to understand our construction.

This construction starts from the generic motive (top of figure). We then apply the linear transformation described by the matrix $\begin{pmatrix} 2 & -1 \\ 1 & 2 \end{pmatrix}$ and display three copies of this transformed zigzag with the beginning notes on the C major triad I_C. Finally, we take the configuration's top notes within C major. They generate the principal theme shown in the bottom score.

12.4 The Moebius Type Motivic Construction

In this section, we want to show how the geometry of nerves of coverings can be used to compose motivic structures in the sonata Allegro movement Op. 3. The nerve of a covering of a set is a geometric representation of the intersections of the covering. The most famous such nerve is the Moebius strip of the covering of a diatonic scale by its seven triadic degree chords. This construction generalizes what Schoenberg called the "harmonic strip". See Fig. 12.4 to understand the construction. We first represent the seven degree chords as points (here small triangles) in space. We then draw lines connecting any two chords that have a non-empty intersection, e.g., II, IV. We then draw a triangular surface for any three degrees having a note in common, e.g., between degrees I, III, V. The resulting surface is a Moebius strip. Its geometry has important consequences for Riemannian harmony, see [118, Ch. 13.4.2.1].

We now describe the motivic construction of a specific part of the composition. Inspired by the harmonic strip of triads, the idea was to construct a covering of the zigzag unit shown on top of Fig. 12.5, which would have a nerve in the shape of a

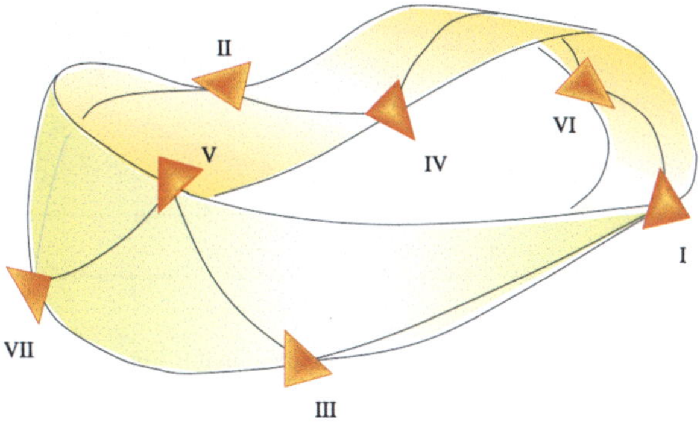

Fig. 12.4 The covering of the major scale by seven three-element degree chords has a nerve that is a Moebius strip. © Guerino Mazzola

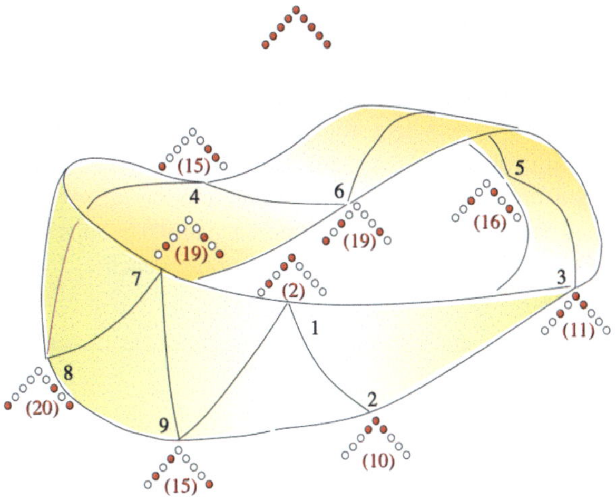

Fig. 12.5 The covering of the motivic unit (top) by nine three-element motives has a nerve that is a Moebius strip. © Guerino Mazzola

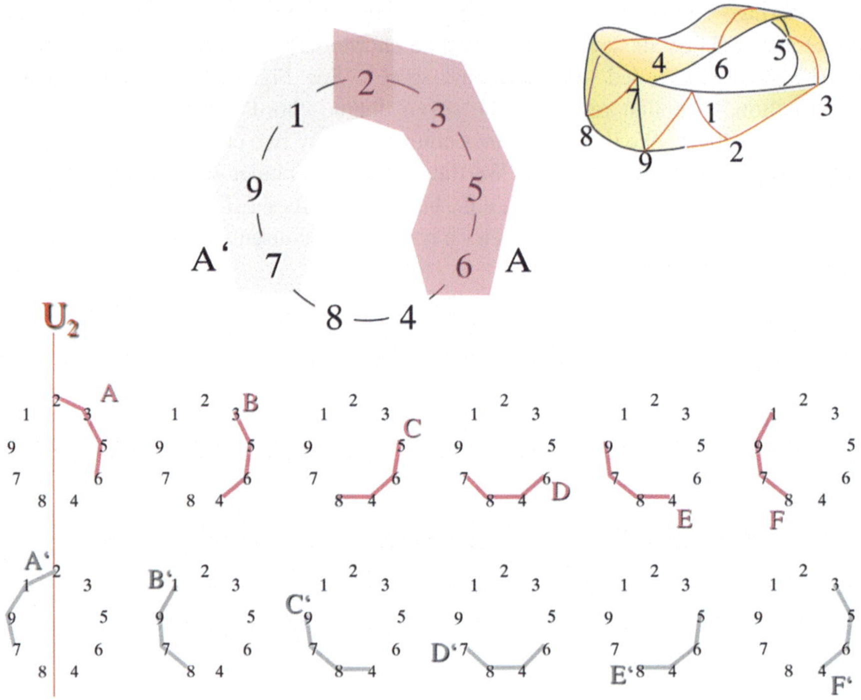

Fig. 12.6 The construction of a concrete melodic structure in this sonata. © Guerino Mazzola

Moebius strip, too. Figure 12.5 shows a solution that consists of nine three-element motives.

Using this scheme of motives, the construction of a concrete melodic structure in this sonata runs as follows, see Fig. 12.6.

We enumerate the nine three-element motives on the Moebius strip according to their path of neighboring motives, yielding the sequence $1, 2, 3, \ldots 7, 9$ of motive numbers. We then select four consecutive motives $A = (2, 3, 5, 6)$ as well as their mirror image $A' = (2, 1, 9, 7)$. Then, the selections A and A' are moved down clockwise and counterclockwise to get the selections B, C, D, E, F and B', C', D', E', F'. We then have two groups A, B, C, D, E, F and A', B', C', D', E', F' of six groups of four motives each.

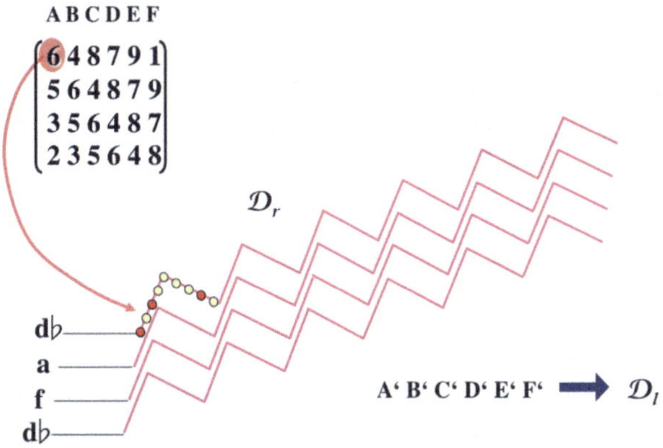

Fig. 12.7 Distribution of the three-element motives in the pitch-onset plane. Group A, B, C, D, E, F yields $\mathcal{D}_r$ (r for right hand), while Group A', B', C', D', E', F' yields $\mathcal{D}_l$ (l for left hand). © Guerino Mazzola

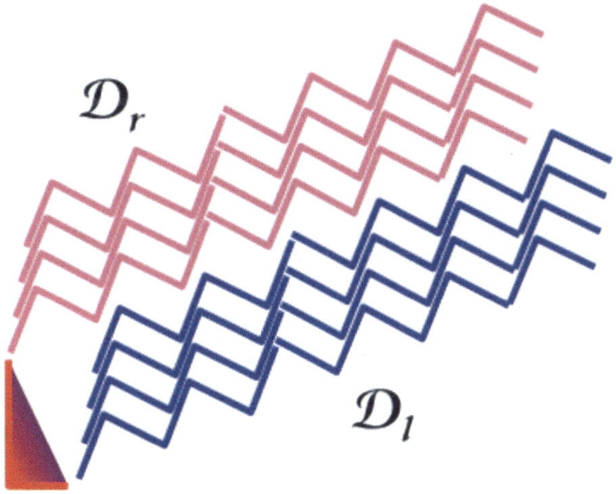

Fig. 12.8 The distribution $\mathcal{D}_r$, $\mathcal{D}_l$. © Guerino Mazzola

These motives are distributed in the pitch and onset time plane as shown in Fig. 12.7 for the group A, B, C, D, E, F, yielding $\mathcal{D}_r$ (r for right hand). The geometric position is given within the original zigzag that is slightly rotated to generate increasing pitches with time. The same geometric positioning is applied for the second group A', B', C', D', E', F' of three-element motives, yielding $\mathcal{D}_l$ (l for left hand). See Fig. 12.8 for this configuration. This produces the measures 33–38 in the sonata, see Fig. 12.9.

Fig. 12.9 Measures 33–38 corresponding to the distribution $\mathcal{D}_r$, $\mathcal{D}_l$. © Guerino Mazzola

12.5 A Model for Future Composition or Just Uncreative Copying?

In view of the question about future developments of musical composition, we should add a short review of the preceding sonata construction.

To begin with, Boulez's idea of a creative analysis is generic and should be a valid model of progress: following Cavaillès' saying that understanding means "catching the gesture and being able of continuing" (see Sect. 14.1.4), Boulez's idea realizes the paradigm of gestural understanding.

A more critical aspect is the mathematical technicality of the preceding approach. Is this technicality eliminating creativity and reducing the composer's activity to uncreative copying? The mathematical language is first of all a *tool of maximal precision* of conceptual manipulation. As such, it cannot be opposed to creativity. Fuzzy approaches are by no means more creative than precise ones. They are only the first sketch of what must be elaborated in the concrete output.

But is the aspect of *calculation* opposed to creativity? One can easily give famous examples, where calculation is an integral part of a compositional method, e.g., Schoenberg's dodecaphonic method, where the calculation of the 48 variants of the row is a basic building block.

In the preceding sonata construction, mathematical calculations were used, but in the same spirit as for Schoenberg: as auxiliary constructions, not as creative pivots. To draw a more down-to-earth image: the compositional machinery needs bolts and gears, and the mathematical calculations are exactly these components, but the design of the machinery as such is in no way reduced to this aspect. Or in a more musical understanding: playing the piano in a creative way must refer to an optimally working mechanical device, but only as a necessary, not a sufficient condition.

One famous argument against mathematical calculation is that they outsource one's creativity to computers as calculation machines, and to reduce one's activity to a "copy and paste" contribution to a musical composition. This argument first confuses mathematics with computer programming, and second, yes, some calculations may be done by computer programs. But these are always those bolts and gears, not more, and not less so.

Nevertheless, the mathematically oriented conceptualization defines a style of thought, and one could argue that such a stylistic determination may limit one's creativity. This argument makes evident the double-edged sword of such a style: on the one hand it limits one's conceptual universe, but on the other it provides us with options that would be unthinkable outside this methodology. The decisive point is that to be creative, one always has be become aware of the walls of one's working box, but see also Chap. 22.

Chapter 13
Imaginary Time: Extending Musical Time Concepts to Cognitive Dimensions

13.1 Einstein's and Hawking's Time Concepts

A dramatic ontological change in modern physics of the twentieth century was the fundamental reconceptualization of time. It was initiated by Albert Einstein's special relativity, where he embedded time in a four-dimensional space-time. Einstein had adopted the space-time approach of Hermann Minkowski, who, in a famous statement, stated that "Henceforth, space for itself, and time for itself shall completely reduce to a mere shadow, and only some sort of union of the two shall preserve independence" [126]. Time then became a multiple variable where the Newtonian singular "divine" time was replaced by a plurality, one for every frame of reference, and different frame times were related to each other by the Lorentz transformation of space-time.

The second revolution of the time concept was primarily introduced by Stephen Hawking [43] and Itzhak Bars [11] in order to solve singularity problems of the evolutionary Big Bang model in the initial moment some 13.8 billion years ago. Hawking's concept of time switches from the real time axis to the plane of complex numbers: Time now has two real coordinates. The complex time $t = t_{\mathbb{R}} + i.t_{Im} \in \mathbb{C}$ pronounces the real time $t_{\mathbb{R}}$ and the imaginary time t_{Im}. This complex ontology has also been proposed and studied by physicists Itzak Bars and John Terning [11].

These two revolutions of the time concept however did not apply to the human cognitive reality. Except in Einstein's case for the popular erroneous statement that "everything is relative". In the case of complex time, Mazzola repeatedly emailed Hawking, but never got an answer, and now, it is too late. He also discussed the issue with other theoretical physicists, but they consistently think of the imaginary component as being a mathematical method, the Wick rotation of real time into imaginary time, not an ontological or even human cognitive topic.

G. Mazzola et al., *The Future of Music*,
https://doi.org/10.1007/978-3-030-39709-8_13

13.2 Musical Consciousness and Creativity

In the following discourse, we will nevertheless argue that complex time could be a key to some of the most virulent problems in the artistic reality of music, namely the nature of artistic consciousness and creativity, especially in the performing perspective of music.

The performing musician, be it for the rendition of a given composition such as a classical score, or for the improvisatory more or less free expression, in jazz for example, faces a complex combination of memory, technique, gestures, and the balance in the famous temporal $\kappa\alpha\iota\rho\acute{o}\zeta$[1] between past and future moments. Successful musical performance is a highly creative and complex activity. At the very center thereof is the sophisticated consciousness of the artist which manages the harmonious collaboration of the above components in real time. "In real time" stands for "in every moment of the performance", in other words: in every infinitesimal point of physical time. This is what every good performer experiences: the complex processual unfolding of musical performance is a rich machinery that defines and is happening in a big space of presence in consciousness. This configuration is described in [111, Ch. 4.12].

This well-known artistic complexification is a miraculous phenomenon since it happens in "too short time", in the physical moment of presence. We are confronted with a big 'space' of consciousness that occupies only a single point in the physical timeline. This makes evident the problematic conceptual status of creative artistic consciousness, and of consciousness in general: How can it be understood that a rich processuality is construed in such little physical time? It can be argued that this type of phenomena is enabled by the existence of a huge 'space' of consciousness that is attached to every moment of physical time. This situation in its acute extremism in performing arts questions a classical rationale for understanding and even defining consciousness in cognitive and neuroscience. If no extra 'space' is added to the classical physical ontology, consciousness cannot be conceived to meet its performative quality. Let us therefore claim the following hypothesis:

Hypothesis 1 *Any workable concept of consciousness, and in particular artistic consciousness in the performing arts, must be construed upon a 'space' that is added to the classical physical spatio-temporal ontology.*

13.3 Descartes' Dualism

The above hypothesis receives a prominent philosophical interpretation if we review the Cartesian dualism which was set up in Descartes' *Principia philosophiae* [26], where he describes the three substances of being: *res extensa, res cogitans*, and God.

---

[1] In Greek philosophy $\kappa\alpha\iota\rho\acute{o}\zeta$ means the right, critical, or opportune moment.

The human existence is comprised of *res extensa* and *res cogitans*, together with their arcane interaction. *Res cogitans* is strongly associated with consciousness. It is not clear how and where this interaction would take place (Descartes' idea of the pineal gland being the crossing locus is too naive). For Descartes a substance is something which can exist without the existence of any other substance, which means in particular that the substance of *res cogitans* is not comprised in the physical ontology of *res extensa* (and vice versa).

The Cartesian dualism is derived from this double substantiality of our human existence, we are 'divided' into the physics of our embodiment and the mentality of consciousness. This complexion is not only related to a double ontology, but more specifically to a double locality: there is a physical as well as a mental space, and both space types are irreducibly separate from each other. Let us be clear about the still persistent problem of dualism that is opposed to the often erroneously propagated reduction of mentality to physics in the neurosciences. The partisans of radical neuroscience claim that ultimately, our thoughts are (however complex) neuronal activities, i.e., thinking is the superficial appearance of physical activities. Such an argument is invalid for the following reason. Suppose we could, for example, explain mathematical thoughts by neuronal activities. Obviously, the description and analysis of such activities would necessarily be made possible by complex mathematical formulas, such as those difficult partial differential equations which describe the axonal transfer of electrical voltage. This means that the explanation of mathematical thoughts would presuppose a sophisticated mathematical machinery, which is a vicious circle: explaining math by use of math does not generate added value. Or, citing the famous *Cahiers* [153, Vol. IV, p. 264] of Paul Valéry: "Imbécile! qui imagine un *cerveau* pour se rechercher l'analyse d'un fait de l'esprit —!"[2]

For these reasons the Cartesian dualism is a second argument for our above hypothesis that consciousness in creative artistic performance has to take place in a 'space' of consciousness that is added to the physical space-time ontology.

But it is evident that understanding this dualism is not the solution of the question about the ontology of consciousness, namely how the mental space is added to the physical reality.

13.4 Synthesizing the Real with the Imaginary: Introducing Complex Time in Music

Our proposal regarding the added mental space of consciousness relates to complex time. The basic idea is that time has a real and an imaginary coordinate, as suggested by theoretical physics. When we consider space-time with complex time, we get a five-dimensional real vector space $ST = \mathbb{R}^3 \oplus \mathbb{C}$ that is the sum of two four-

[2]Imbecile! who imagines a brain to search for the analysis of a fact of the mind!

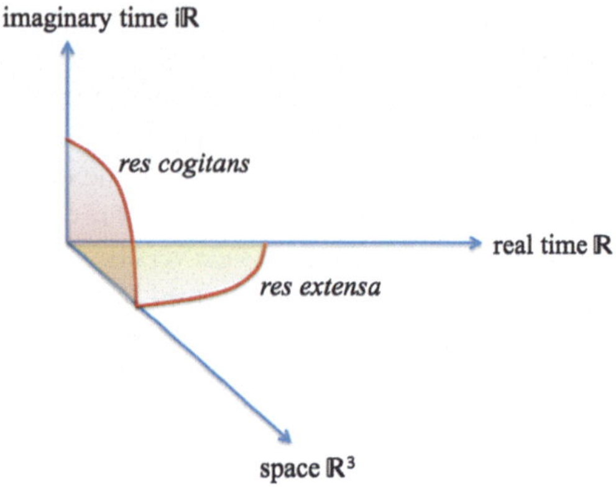

Fig. 13.1 Descartes' dualism vanishes in a space-time with complex time. © Guerino Mazzola

dimensional subpaces, $ST = RST + IST = \mathbb{R}^3 \oplus \mathbb{R} + \mathbb{R}^3 \oplus i\mathbb{R}$, the physical space-time RST and the mental space-time IST, see Fig. 13.1.

The physical and mental subspaces have the spatial part in common: $RST \cap IST = \mathbb{R}^3$. This connection would replace the Cartesian function of the pineal gland. Our configuration separates *res extensa* from *res cogitans*, but it also gives them a shared subspace. This approach means that we are proposing a conceptual synthesis of the Cartesian divide. *Res cogitans* and *res extensa* are two subspaces of a big space, defined by two components of complex space-time. It is remarkable that Jonathan D. Kramer in his important book *the Time of Music* [99, p. 55] proposes a "vertical time"—similar to the geometric orthogonality of our imaginary time, but without any reference to complex time concepts—that is characterized by: "The result is a single present stretched out into an enormous duration, a potentially infinite "now" that nonetheless feels like an instant."

The next step must be a theory of interaction of these four-dimensional subspaces which comprises the laws of physics and the dynamics of consciousness. To get off the ground, we have initiated research that models musical performance as a transformation of symbolic reality (of a score for example) to physical reality, see [113].

13.5 Performing Symbols to Physical Gestures

Our previous performance theory dealt with the transformation of note symbols, as given in a common score, to physical (acoustical) events. This is similar to the approach taken by contemporary performance theory as initiated by the school of Johan Sundberg in Stockholm. Our theory used standard methods of differential geometry. It was one of the most important results of our research that the refinement process of performance involves Lie derivatives of performance vector fields, the tempo curve being the classical example of such a performance field.

In view of the union of physical and symbolic/mental realities as suggested by the above *ST* space-time, it became imperative to also view the performance process more embodied than by abstract note symbols and their acoustical realization in performance. In [113], we set forth an embodied performance theory which instead of note sequences deals with musical gestures, configurations of curves in space-time that may be realized by the musician's movements of human body limbs, the vocal tract, arms, and hands.

This approach is paralleled by the elegant physical theory of strings, which does not describe elementary objects as points that move in space-time, but as parametrized curves which move through space-time and thereby trace a "world-sheet" (instead of the well-known "world-line" of traditional particle dynamics). In our musical performance theory, the string would correspond to a musical gesture, for example an up-down movement of a pianist's finger. Its world-sheet would be the surface that is spanned between the symbolic finger movement as described on the score and the physical movement in performance, where a real physical gesture is happening. Figure 13.2 shows such a world-sheet. The blue rectangular line to the left is the symbolic gesture, while the red curve to the right is the physical realization the symbolic gesture.

The performance theory of musical gesture strings parallels the physical theory in that the shape of such a world-sheet is determined by the minimization of a Lagrange potential that is associated with a world-sheet, see [113] for details.

In this model, the symbolic gesture evolves in imaginary time of consciousness, while the physical one evolves in physical time. This means that we have a world-sheet of time from the symbolic/mental time line to the physical one. This world-sheet turns out to be a conformal mapping in complex space. Figure 13.3 shows the

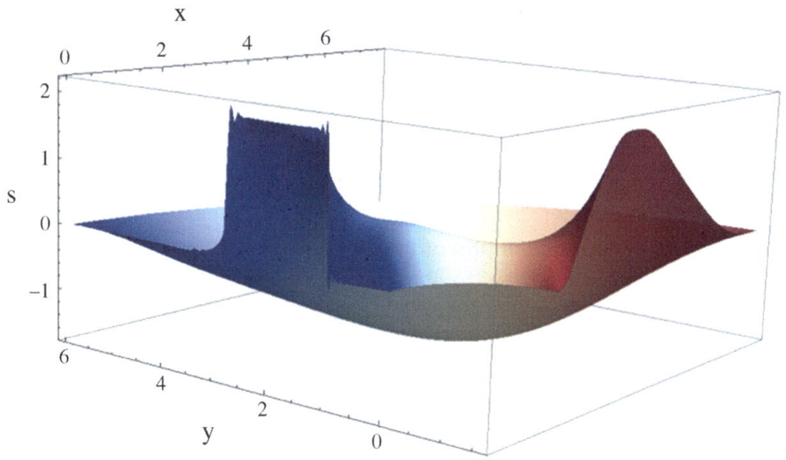

Fig. 13.2 World-sheet of the passage from symbols (blue) a physical up-down movement of a pianist's finger. © Guerino Mazzola

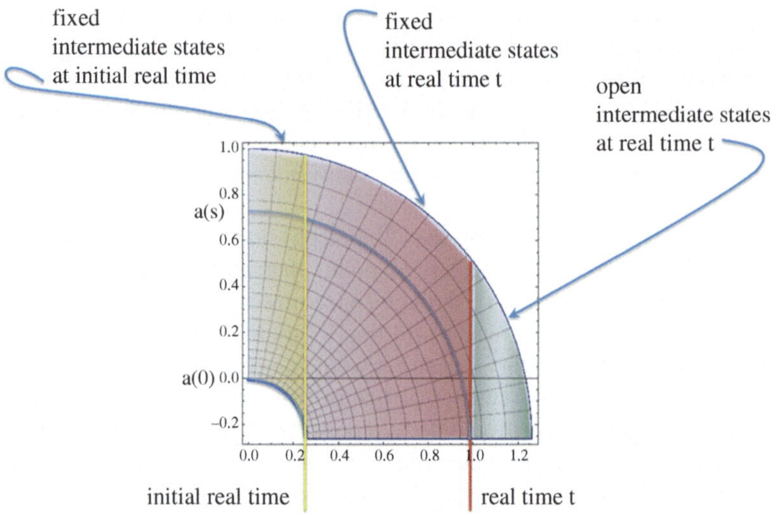

Fig. 13.3 The world-sheet of complex time of a gesture and its intermediate states. © Guerino Mazzola

temporal world-sheet and its intermediate levels when understood as a function of real/physical time.

Here we face a completely new research in consciousness when morphing from mental to physical time.

An important new results of this theory is perfectly parallel to the above mentioned result of classical performance theory about role of the Lie derivative in the refinement process of performance vector fields. Again, in the gestural setup, there is an operator of Lie derivative type that captures refinements of gestural performance.

13.6 Application of Imaginary Time to Composition

Let us make two concrete examples of composition relating to the imaginary time concept. The composer will have to imagine time when writing a score, this is not the physical time of performance but an imagined fictitious time, which we identify with the imaginary time. In the composer's thoughts this time should help shape the succession of the score's notes. This imagined gesture would then be transformed

Fig. 13.4 Beethoven Sonata F minor Op. 57, first movement, measures 14–46 extracted from: Ludwig van Beethoven, Sonaten für Klavier, Band 2, Wiener Urtext Edition, UT 50108, Vienna 1999. Image reprinted with kind permission from Wiener Urtext Edition

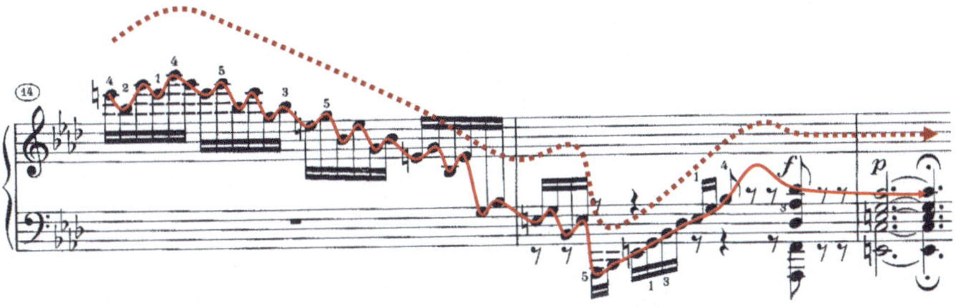

Fig. 13.5 Big descending gesture in an imaginary time line. Beethoven Sonata F minor Op. 57, first movement, measures 14–46 extracted from: Ludwig van Beethoven, Sonaten für Klavier, Band 2, Wiener Urtext Edition, UT 50108, Vienna 1999. Image reprinted with kind permission from Wiener Urtext Edition

into a real gesture, where the performer's limbs move in physical time. If this transformation is not thought of, the composition might turn out to be "unplayable" or at least opposed to the performer's embodied capabilities.

A first example is Beethoven *Piano Sonata Op. 57*, first movement, big descending gesture in measures 14–16, see Fig. 13.4. Here the composer has imagined a very pianistic gesture, which we show in Fig. 13.5. Not only does this induce a perfect transformation in the above sense, it is also bad performance not to recognize this imaginary gesture as a model for performance, an error which Glenn Gould committed in his famously wrong performance of *Op. 57* [40].

The second example is Boulez's *Structures pour deux pianos* II, chapitre I, p. 17, part 80, see Fig. 13.6. This serial composition is a reworked version of the

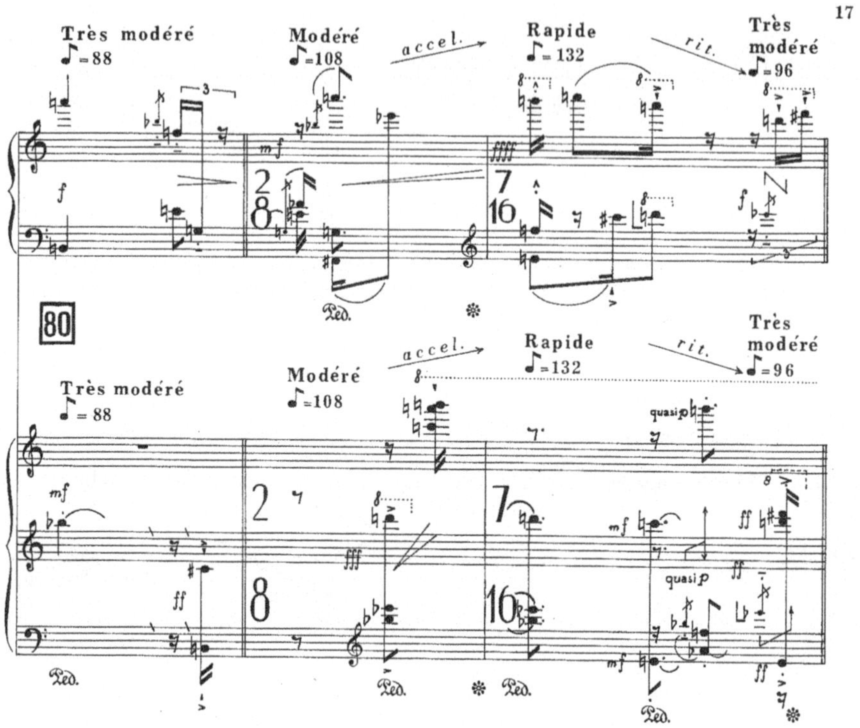

Fig. 13.6 Boulez, *Structures pour deux pianos* II, chapitre I, p. 17, part 80. © 1967 by Universal Edition (London) Ltd., London/UE13833

composer's *Structures pour deux pianos I*, which we discussed in Chap. 11. It is evident from the serial principles and visible from the score that gestural aspects are not realized in imaginary time, and it is therefore very difficult for the two pianists to move their limbs in a pianistic way.

Chapter 14
Mathematical Gesture Theory

The most famous gesture, 'the' gesture, is probably the one of God's Creation of Adam, in the Michelangelo's well-known fresco in Cappella Sistina.

The importance of gestures in music is crucial: performers learn specific gestures to play musical instruments, and variations of these gestures allow different musical-acoustic results. These are the changes of loudness, tempo, and pitch. The importance of gestures can also be gleaned from their historical background, as we will see in the following section.

Western musical indexnotationnotation developed from *neumes*, a system which fixes the shape of the melody and the movements of the choir conductor's hands (chironomic notation) to paper (Fig. 14.1). It is interesting to note that, in a culturally and geographically different environment, musical notation had a similar gestural origin. We have already explored the case of Gregorian music for Western culture. For the Eastern cultures, early Chinese music notation was based on gestures. The notation represented gestural instructions (production) for the player. These were largely created for an instrument called the Guqin (古琴).

14.1 Historical Roots

Summary In this section, we will discuss the thoughts of philosophers who emphasized the importance of gestures.

14.1.1 Tommaso Campanella

Tommaso Campanella (1568–1639) was a Dominican friar and theologian who relied heavily on the importance of the senses in his writings. From his text,

G. Mazzola et al., *The Future of Music*,
https://doi.org/10.1007/978-3-030-39709-8_14

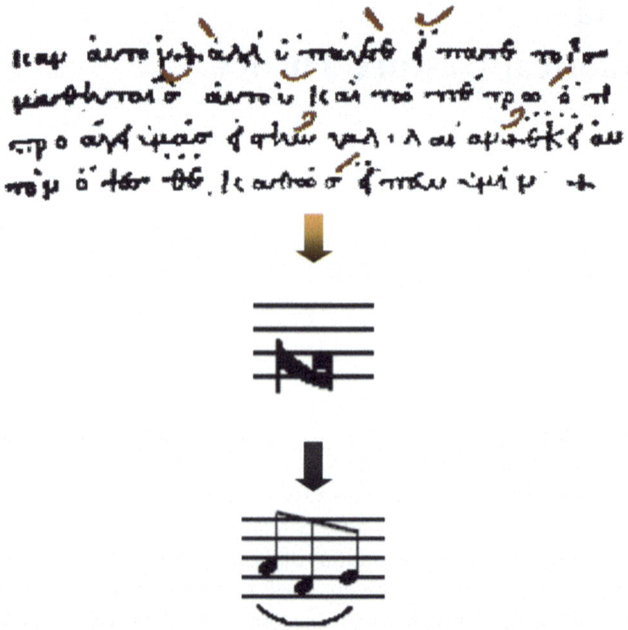

Fig. 14.1 Western musical notation derives from Gregorian notational system. © Guerino Mazzola

whose title can be roughly translated to rational philosophy, he discussed five parts: *dialectica, grammatica, rhetorica, poetica, and historiographia* [20, 144]. His primary message as it pertains to creative gestures is as follows:

> . . . *omnes propositiones per singulares tamquam ad digitum exponuntur.*
> . . .pointing with your finger is the only certitude.

Campanella simply declares that gestures are the only thing of which we can be completely sure. Gestures do not rely on a byproduct to prove their existence, for a rotation is just a rotation. There may be strokes of paint, but that is only a visible manifestation of the gesture which had to have occurred. Campanella's words speak to the truth that gestures are universal.

14.1.2 Hugues de Saint-Victor

Hugues de Saint-Victor (c. 1096–1141) was a Saxon theologian. He gives a most adequate definition of a gesture. He describes it as something which is multi-faceted and has factors which must be realized across different realities [145].

> *Gestus est motus et figuratio membrorum corporis,*
> *ad omnem agendi et habendi modum.*

> Gesture is the movement and figuration of the body's limbs with an aim,
> but also according to the measure and modality
> proper to the achievement of all action and attitude.

Hugues de Saint-Victor declared that gestures have not only motion, but also an attitude. This is essential to the importance of gestures as a concept. Although gestural motion can be reduced to a mathematical formula, the formula will abandon key information which the original gesture held innately. This means that to fully understand artists and their decisions, we must interpret their works at the gestural level.

14.1.3 Paul Valéry

Paul Valéry (1871–1945) was a French poet and philosopher. He was nominated for the Nobel Prize in Literature on 12 separate occasions. In addition to his artistic expertise, he fostered a special interest in optics. His words on art translate well to mathematics and science.

> *C'est l'execution du poème qui est le poème.*
> It's the rendition of a poem which is the poem.
> (The interest of science lies in the art of *doing* science.)

Valéry's philosophy was integral to his pursuits in science. However, he defined a separation between an artistic product (such as poem, painting, or song) and the actual artistic making. To him, the art was not the physical byproduct. The art was in the doing and creating, the gesture. He brought this approach to his scientific studies. He believed that the joy of science was not only in the answers that were found, but even more in the questions that were asked.

14.1.4 Jean Cavaillès

Jean Cavaillès (1903–1944) was a French philosopher and mathematician who was particularly interested in the philosophy of science. His words surrounding the role of gestures [21] have had an incredible impact in the direction of this book.

> *Comprendre est attraper le geste et pouvoir continuer.*
> Understanding is catching the gesture and being capable of continuing.

Cavaillès' words are the entire foundation of this topic of gestures; they apply to every aspect of the arts. Imagine that you are an improvising musician in a jazz combo. To play something which involves skill would be to play based upon the context which you receive from your fellow musicians. Of course, there is still room for personal flair. However, it only makes sense that you would generally play in the context which the environment decided.

Imagine that you are a solo painter. To convey a coherent message through your work, your gestures must be linked somehow. Even Jackson Pollock's gestures have visible connections between them. Just as an orchestra follows the gestures of their conductor, so does an audience interpret the gestures of a painting.

14.1.5 Maurice Merleau-Ponty

The French philosopher Maurice Merleau-Ponty (1908–1961) beautifully summarized the importance of gesture within communication. His illustrated the symbiotic relationship between the communication of ideas and gestures [123].

> *La parole est un vèritable geste et elle contient son sens comme le geste contient le sien. C'est ce qui rend possible la communication.*
>
> Language is a veritable gesture and it contains its sense much as the gesture contains its own. This is what makes communication possible.

Merleau-Ponty combined the ideas of Cavaillès and de Saint-Victor. He proposed that a gesture contains information within itself. Connecting these gestures is like stringing together the words of a sentence. Therefore, we are able to create deeper meaning when we link more gestures together. This brings us to the topic of defining gestures and hypergestures.

14.2 Definition of a Gesture

Besides the fact that the use and the concept of gesture is particularly relevant to music, its precise definition is very recent. The mathematical definition of gesture, given by Mazzola [115], is the following. Let's think of a dancer. He or she touches the stage in discrete points, while moving continuously. If we compare the points to the musical notes, and the continuous curves to the gestures, we can understand their importance for musical performance! And also for composition, because composers also think of final gestures while choosing and writing specific symbols on the musical staff.

Here you are. Let's follow the graphic representation in Fig. 14.2. To the left, we have an abstract structure (music? painting? cooking? nothing yet) of points and connecting arrows, called *skeleton*.[1] To the right, we have a space with the keyboard of the piano indicating the change of position of the hand along this direction to pick a specific pitch, the position above the keyboard (for the 'states' pressed-not pressed key), and the time, the onset. The abstract skeleton is mapped into a system of continuous curves connecting points in space and time, the gesture's *body*. It is

[1] Points and arrows are the starting spot of the mathematical theory of categories, that has been applied to music by Mazzola [118].

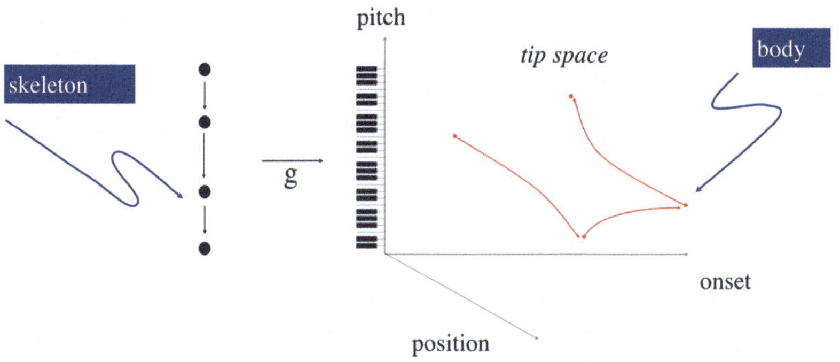

Fig. 14.2 Illustration of the mathematical definition of a gesture, as a mapping from an abstract structure, a skeleton of points and arrows, to a system of continuous curves in space and time, with the same number of points and connecting arrows. The example shown is about the pianistic gesture. © Guerino Mazzola

clear that we can define gesture for other situations while changing the names of the axis for the space on the right. The number of points and arrows, of course, must be the same. A such mapping is a *gesture*.

The transformation from notes to sounds is described by the mathematical performance theory of notes [118], from the written information inside the score to the reality of sounds of acoustics. The transformation from gestures 'suggested' by the score into the real movements made by performers, is object of ongoing research, and we already have the first results [105].

How can we get gestures from the score? Let's consider the case of a pianist. If we look literally at the content of a score, we get a series of instructions such as 'play this note at this time with this loudness.' This is more or less what happens for a MIDI instrument! And the change of position should happen *instantaneously*. Not even the world's fastest pianist can play with an infinite speed, because it is not allowed by the laws of physics. Sorry for that.

What happens in reality is that the performers need a *finite* time to make a movement. We call symbolic gestures the curve systems of the first type; physical gestures those of the second one. The surface that transforms a gesture of the first type into one of the second is called a world-sheet,. The name and the way to obtain it (via math-theoretical physical methods) comes from String Theory in theoretical physics. Again, another interdisciplinary parallelism: elementary particles in physics are no longer seen as 'points' but as vibrating strings; in a similar way, music can be described not only in term of 'points' but also in terms of gestures. This world-sheet surface is also a hypergesture. We will define it in Sect. 14.3.

The shape of the surface depends on the choice of the force field that shapes the surface. We will not deal with mathematical details here, but we can say that the force field plays the role of the operators that transform the notes of a score into the sounds [105, 118]. Moreover, the time flowing inside the score and the time of the physical performance are qualitatively different: they represent a simplification of time of mental reality versus physical time, see also Chap. 13 for a mathematical description of this approach.

14.3 Hypergestures

In her book about piano performance, Renate Wieland wrote: "Die Klangberührung ist das Ziel der zusammenfassenden Geste, der Anschlag ist sozusagen die Geste in der Geste." (The sound contact is the target of the embracing gesture, the touch is so to speak the gesture within the gesture). It refers to a more complex concept of gestural utterances. We can make this idea precise in the following sense.

A gesture connects points in space and time. If a gesture is seen as a point itself, the gesture that connects such gesture-points is called *hypergesture*. A hypergesture is a gesture of gestures. Figure 14.3 shows a surface obtained as a loop moving within space along a circle-path.

The concept of hypergesture is very powerful. It helps create incredibly complicated structures of *nested* gestures. An entire symphony, for example, can be studied in terms of hypergestures. With a hierarchical structure, we can describe from top to down the gestures inside an symphonic orchestra. We can think of all the instrumental sections, all the performers of each section, all the micro-movements of the left hand of each violinist This is also a hypergesture. The gestures of each level are composed of smaller gestures of next level.

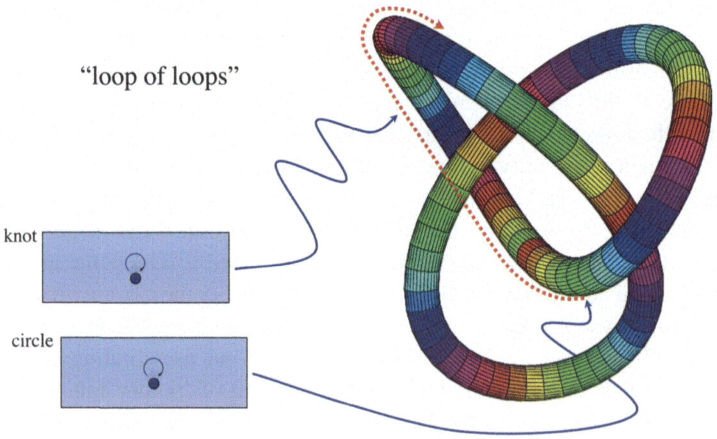

Fig. 14.3 A surface obtained as a loop of loops to represent the concept of hypergesture. © Guerino Mazzola

14.4 Future Technology for Gestures

The mathematical theory of gestures is also a basis for future technologies, which represent and communicate the realm of (hyper)gestures. A language that would replace/enhance MIDI must be conceived, and it seems that the mobile network standard 5G, also significantly called the "internet of skills", might be a technological media for gestures.

Chapter 15
Future Theories (Counterpoint Etc.)

In this chapter we first give examples of (potential) future music theories and then discuss their role for creativity in musical composition. But let us first have a look at the role of theories to this date. In the very beginning of music as a knowledge space in Europe, the Pythagorean school conceived music as an experimental tool (practiced on the lyra or monochord) to connect to the "world formula" tetractys. Music was just the experimental but not the artistic autonomous practice. This attitude changed in the Middle Ages where the individual utterance of humans as opposed to divine preconception became a new understanding of the arts. The development of counterpoint and the development of the central perspective are the outcome of this existential shift that flourished in the Renaissance.

Music theory then changed its role from the transcendental and a priori Pythagorean truth to an intellectual device that explains *ex post* what artists are doing driven by their own individuality. Descartes added to this development the psychological rationale for musical expression, replacing mathematical rationality by emotional forces and effects. In the eighteenth and nineteenth centuries, some mathematical aspects of music theory reentered the field, especially through Leonhard Euler's geometric representation of the pitch space and later Hugo Riemann's musical logic, especially his idea about tonality to be an abstract function on the set of chords.

In the first decade of the twentieth century the ample extension of harmonic languages created by composers, such as Beethoven, Liszt, Busoni, or Mahler, lead to a scission between a theory that was too weak to explain or drive new developments and composers who were seeking for new perspectives. The disruption from this broken tradition was—after a phase of "atonal experiments"—concluded in 1921 by Schoenberg's proposition of a compositional method based on dodecaphonic rows, the twelve-tone method. We discuss this method in Chap. 11. This approach was however not a theory, but a theoretically insignificant recipe for composing music. We make this clear in Sect. 11.4.

G. Mazzola et al., *The Future of Music*,
https://doi.org/10.1007/978-3-030-39709-8_15

The twentieth century was therefore given a strong compositional vector, also including new attempts to extended tonal or rhythmic conceptions by composers such as Scriabin, Bartók, Debussy, Ravel, Prokofiev, or Ligeti. The development of music theories was revitalized by new mathematical methods (America Set Theory, Wolfgang Graeser's symmetry theory, and the European Mathematical Music Theory, the latter being described in Chap. 10), together with music theory software for metric, motivic, or harmonic analysis (see Sect. 6.3).

This new development unfolded in two directions: On the one hand, broken music theories, especially harmony and counterpoint, could be repaired and extended, the mathematical theory of counterpoint is discussed in Sect. 15.1. Mathematical harmony is applied to the composition of a sonata, which we describe in Chap. 12. These new directions are however conceptually different from the theories of the past, which only explained already created works. They have the tendency of resembling more and more physical theories in that not only past, but also future events can be described: *the explanatory mode is enriched by a predictive one*. Of course, composers cannot be forced to follow a given theory, but the development is more subtle: To begin with, new theories tend to enlarge the conceptual framework for composition, they offer new musical perspectives. Perspectives upon musical landscapes that have not been explored yet. But more than dodecaphonic composition, these new perspectives are theoretically grounded. And they are also accessible to computer-supported compositional tools. One such tool is described in Sect. 17.2.3.

It is important to understand this new trend of musical composition since we are far from having composed everything. Not even in established styles. Why would it be unattractive to compose a piano sonata in Beethoven's style? *Op. 112*? Or also mixing different historical style episodes into one specific composition? The historicity of music here becomes a virulent topic that is enhanced by new methods and tools. See our discussion of historicity in Chap. 12.

The following section discusses a future counterpoint theory. We then conclude this chapter with a confrontation of the idea of future theories with musical creativity: are they in conflict or can we think of future theories as catalyzers of creativity?

15.1 Future Counterpoint Theories

The development of counterpoint is not only a musical topic but rather a fascinating and creative switch between musical and mathematical ontologies [112]. Such a switch seems to be of increasing relevance in view of the strong implementation of mathematical methods in computer technology.

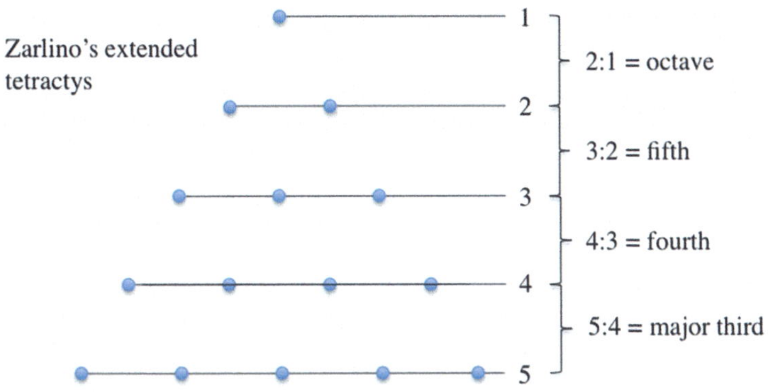

Fig. 15.1 Zarlino's pentractys. It includes the Pythagorean tetractys and extends it by adding the number five to represent the major third. © Guerino Mazzola

15.1.1 The History of Counterpoint Until Palestrina

Counterpoint deals with the collaboration of two or more voices under the perspective of interval qualities called consonance and dissonance. The history of its development is not only a musical one but one of changing musical ontology to its mathematical counterpart and also one of a first experimental paradigm long before Galilei introduced experiments in the European creation of physics.

15.1.1.1 Pythagoras and His Heritage

The Pythagorean origin of musical interval concepts is based upon his *tetractys*, the triangular arrangement of ten points, a metaphysical "world formula" showing on top one point, followed by two, three, and four points, see Fig. 15.1.

This formula could be listened to by playing the monochord, a kind of early experimental device to access the world formula. The ratio 1:2 was for the musical octave, following the ratio 2:3 for the perfect fifth, and then 3:4 for the perfect fourth. These intervals were the original consonant intervals of early European music. Their simple mathematical rationale was the theoretical justification for their quality. During the Middle Ages until 900 AD, these intervals were considered valid consonances, together with some mathematically constructed additional intervals.

15.1.1.2 Seven Hundred Years of Experiments

Around 900 AD, the composition of several simultaneous voices, added to the Gregorian chant, initiated an experimental phase of 700 years, where the Pythagorean principle was no longer stable. Initially, two voices singing in parallels of fourth or fifth intervals were composed, but successively, composers started experimenting with other consonant or dissonant intervals. Sometimes, sixths were thought to be consonant, then they changed to be dissonant. Successively, the fourth also was put into question and eventually became dissonant. When in the sixteenth century Giovanni da Palestrina created his polyphonic works, the experimentation was virtually terminated and we see the final concept of a consonance, which has been canonized in Johann Joseph Fux's *Gradus ad parnassum* in 1725 [38], where within an octave, the prime, minor and major thirds, perfect fifth, and minor and major sixths are consonant. The other six intervals, minor and major second, fourth, tritone, minor and major seventh are dissonant. In modern terms, speaking about intervals as numbers between 0 and 11, the set of consonances is $K = \{0, 3, 4, 7, 8, 9\}$, while the dissonances are $D = \{1, 2, 5, 6, 10, 11\}$.

15.1.2 The Miraculous Effect of Composition on Consonances and Dissonances

It is not clear to this date, what was the *movens* of this 700 years' experimentation, and in particular, why the consonant Pythagorean fourth with the simple frequency ratio of 3:4 turned out to be dissonant, whereas the equally simple ratio of 2:3 of the fifth kept this interval within the set of consonances. This is all the more astonishing since by then, the major third frequency ratio 4:5 was accepted as a consonance. So why would one take away the even simpler ratio 3:4? The prominent music theorist Carl Dahlhaus has argued that the reason for this result of the long experimentation was not an arithmetic, psychological or physiological one but one following from rationales of musical composition. But he could not identify which rationale could have caused this result.

15.1.2.1 The Dissonant Fourth and Forbidden Parallels of Fifths

Summarizing, the stable result of contrapuntal systematization after a long period of experimental investigations had two mysterious characteristics: the dissonant fourth and the forbidden parallels of fifths. This is remarkable also because the dominant Pythagorean ideology of the tetractys (and its extension by Zarlino to the ratio 4:5) was still accepted as a fundamental law of metaphysical ratios.

15.1.3 The Mathematical Understanding of the Miracle

What had effectively happened during the 700 years of experimentation was a subtle shift of the very concept of a consonance or dissonance. In the Pythagorean perspective, deciding whether an interval was consonant or dissonant was a matter of inspecting the frequency ratio (or, from a more practical point of view: the ratio of the string subdivision on the monochord) of the individual interval. This point of view was however not a musical one for the working composer. A composer is more interested in the contextual behavior of intervals, not in their properties as isolated individuals. This means that the classification of intervals was more a question of how intervals behave in the community of other intervals. In this understanding, an interval is not consonant because it has an attribute (such as a simple frequency ratio) as an individual, but because it is a member of a collection of intervals that behave well together. Replacing the Pythagorean attribute of simple frequency ratios one now would rather have to ask what defines a consonance *as a member of a collection of intervals*.

15.1.3.1 The Symmetry Between Consonances and Dissonances

This musically motivated shift of the concept of consonances vs. dissonances could be explained by a mathematical insight [108, Ch. 7], namely the fact that there is a unique symmetry $T^2 5 : K \overset{\sim}{\to} D$ that maps consonances to dissonances. And the same $T^2 5$ also maps dissonances to consonances. This so-called autocomplementarity symmetry is the basic fact which enables the modeling of Fuxian counterpoint [118, Part VII]. In particular, the forbidden parallels of fifths are deduced without any psychological explanation in this model. The model is using deformation symmetries, similar to elastic forces that act upon the dichotomy of consonances and dissonances.

15.1.3.2 Five New Worlds, Raga, and Scriabin

What about the dissonant fourth? If one classifies all dichotomies, i.e., splittings of the 12 interval numbers in $\mathbb{Z}_{12}$ into two disjoint six-element subsets X, Y, then there are only six classes of such dichotomies which admit a unique autocomplementarity. Here are the six representatives, together with their autocomplementarity:

$$\#64 : (\{2,4,5,7,9,11\}, \{0,1,3,6,8,10\}), AC = T^5.11$$

$$\#68 : (\{0,1,2,3,5,8\}, \{4,6,7,9,10,11\}), AC = T^6.5$$

$$\#71 : (\{0,1,2,3,6,7\}, \{4,5,8,9,10,11\}), AC = T^{11}.11$$

$$\#75 : (\{0,1,2,4,5,8\}, \{3,6,7,9,10,11\}), AC = T^{11}.11$$

$$\#78 : (\{0,1,2,4,6,10\}, \{3,5,7,8,9,11\}), AC = T^9.11$$

$$\#82 : (\{0,3,4,7,8,9\}, \{1,2,5,6,10,11\}), AC = T^2.5$$

We call such dichotomies *strong*. The left number is the label given in the classification theory of subsets of $\mathbb{Z}_{12}$, see [118, Ch. L.1]. The Fux dichotomy

(K/D) is number 82. It can be shown that this dichotomy has a geometric distribution of its two halves that is distinguished from the other five strong dichotomies. It is characterized by purely geometric properties, meaning that the Fux dichotomy which was achieved in Palestrina's epoch is defined not only by the experimental work of many composers, but more mathematically by a purely geometric property, see [118, Ch. 30.1] for details. In the context of Depth EEG investigations in humans, especially in the hippocampal formation of the emotional brain, it could be shown that the EEG for Fuxian consonances was significantly different from the EEG for Fuxian dissonances [118, Ch. 30.2].

The model for Fuxian counterpoint can also be applied to any of the other five strong dichotomies. This means that now, we have six contrapuntal "worlds" instead of only one. In [7], these new perspectives have been developed, including software for the creation of counterpoint in any of these worlds. Counterpoint for class 78 has been used in a composition, *Black Summer*, by Joomi Park, see [117]. Figure 15.2 shows some of her "consonant" intervals of class 78. Observe that the unison is dissonant here and does not play a role of a relaxed moment.

Apart from the possibility of creating new compositions in different contrapuntal worlds, it is also interesting that to understand difficult constructions in Alexander Scriabin's sonatas, it could be shown [120, Ch. 21.2.3] that the strong dichotomy 78 is realized with Scriabin's famous mystic chord (its six pitches define one half of dichotomy 78). It can also be shown [6] that the contrapuntal world of this dichotomy is relevant in Scriabin's counterpoint, e.g., in Sonata No. 5.

Moreover, dichotomy 64 is an interesting structure related to the diatonic scale: its first half consists of the six proper intervals in C major from the tonic: $C - D = 2, C - E = 4, \ldots C - B = 11$. Now, the C major scale is optimal for counterpoint in Fux's world. For counterpoint in world 64, the best scale is a scale which pertains

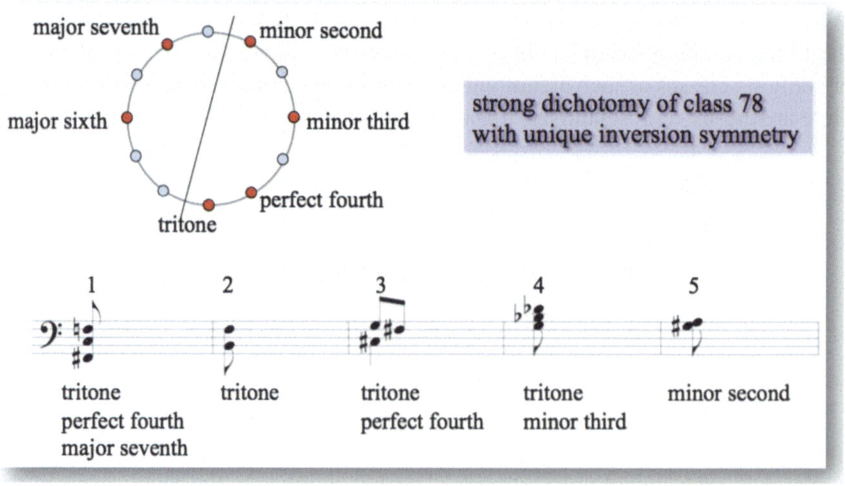

Fig. 15.2 Some consonant intervals in Park's composition. © Guerino Mazzola

to Raga music: $K^* = \{0, 3, 4, 7, 8, 9, 11\}$. This scale is very similar to the famous mela "mayamalavagaula" that can be represented by $\{0, 3, 4, 7, 8, 9, 1\}$. This latter scale is very similar to the Fux consonance part. So we have a funny switch: the major scale is good for Fux, and the "consonance" mela is good for the major dichotomy No. 64. We do however not know how deeply the counterpoint world of this dichotomy is engrained in Raga music.

15.1.3.3 Counterpoint for Microtonal Tuning

A major problem for the future of counterpoint is a possible setup for microtonal scales. The traditional concept of consonances is no longer applicable for microtonal situations. Here the mathematical model was successfully applied by Octavio Agustín-Aquino to extend counterpoint to arbitrarily refined microtonal scales, see [7] for details. This is a remarkable result that proves the power of generalization of mathematical models. Despite the impossibility to define in a reasonable way consonances in microtonal environments, the mathematical symmetries work perfectly here to create contrapuntal structures in exotic tunings.

15.1.4 Future Contrapuntal Perspectives

Besides the opening of five new contrapuntal worlds, we also have to think of connecting these worlds: How can we morph a composition by Johann Sebastian Bach in the world of Fux (with some exceptions accepted) to a composition in the world of the mystic chord, or in the world of the major dichotomy No. 64? Some first ideas are presented in [7], but we need an experimental environment for new directions in contrapuntal creativity. Can we hope that good technology may help find interesting avenues, perhaps also supported by new instruments/sound colors and powerful algorithms to deal with microtonal approaches?

A central question in this field should deal with the problem of how to educate young composers to question the one and only Fuxian tradition and to seek for new frontiers in worlds of unheard polyphonic richness, and also transcending the rigid Western canon.

15.2 Future Theories and Creativity

The fundamental question here, and more generally everywhere when new tools are introduced that may interact with creativity, is about whether creativity is restricted or enhanced. Let us make clear the difference between past situations and the present one.

Earlier, theory was a structural device for better conceiving the composer's ideas. For example, it could help shape a tonal modulation according to an established harmonic language with its concepts (functional determinants, cadences, pivotal chords, etc.) and theorems (e.g., a selection of advantageous chord combinations as a function of the modulatory tonalities).

But now future theories, which are theories of the potential future of musical composition, not only describe language and theorems for compositional ideas, but provide the composer with the ideas as such. The future counterpoint theory as described above opens new ways of conceiving interval qualities and their relationships, not only for the classical 12-tempered chromatic scale, but for microtonal environments where no intervallic experience is at hand. This type of creative environment is challenging not only technically, but also with respect to an absent experience, one cannot compare one's creation with similar works. Contrasting to serialism, there is a strong conceptual connection to past theories and methods, one knows that the future theory is intellectually valid as a logical extension of given approaches. But is this enough to guarantee interesting future compositions?

Therefore a first problem rises: Can we trust in the musical logic when exploring new grounds? The fundamental question stated above then becomes more precise. Do we have to follow musical logic when being creative? Is creativity not the opposite of following a given thread? We come back to this question in Chap. 22, which deals with creativity.

Part IV
Cultural Extensions

Chapter 16
A Critique of the Western Concept of Music

16.1 Disembodied Music

Due to the Christian tradition, the human body is the main frame of our sinful origin. The development of European music in the Middle Ages took place within the expansion of the Christian religion and can be characterized by the corresponding ideology of adoring God's eternal disembodied perspective *sub specie aeternitatis* (under the aspect of eternity). Musical creation was understood and developed as the fabrication of an image of eternal laws of divine proportions of Pythagorean consonances (octave 2/1, fifth 3/2, fourth 4/3, see Sect. 15.1.1.1). The human individual, and more concretely, the body of the musician, were necessary, but more disturbing than enriching components of musical practice. The complete control of one's body, including the utmost reduction of its inner dynamics, sweat, limbic expressivity, and dancing movements, was and often still is a *conditio sine qua non* (a necessary condition) for a cultivated European approach to music. The free jazz piano titan Cecil Taylor accordingly stated: "In Western music, the body plays no role."

16.2 Absent Gesture Theory

Despite the fact that the historic development of European musical notation originated in the neumes, which were gestural symbols, gestures have not been developed in the context of Western music theory. Gestures are well known to performers, but only as practical, "applied" functions to communicate the score's symbolic truths. It is remarkable that gestures were discussed by prominent theorists, such as Theodor Wiesengrund Adorno [2] or Roger Sessions [149]. Also in piano performance theory, Renate Wieland and Jürgen Uhde [157] have discussed gestures. But none of these discourses have ever been transformed into a music theory of gestures that

G. Mazzola et al., *The Future of Music*,
https://doi.org/10.1007/978-3-030-39709-8_16

would be comparable to harmony or counterpoint. A theory of musical gestures has only been developed recently by Mazzola and collaborators [111, 113, 116]. In non-Western music cultures, gestures are sometimes core concepts. The Japanese Noh theater theory is an excellent example, see Sect. 21.2.2.

16.3 Paper Music Fiction

As a consequence of the disembodied culture of music, the score notation developed into a central representation of music, its performance being a physical realization that only maps the eternal abstract truth of the scored musical essence into sensual contingency. The situation is parallel to mathematics, where, for example, drawing a triangle is only a secondary representation of an abstract conceptual entity. The musical thought is defined in the score, performance is not adding any substantial information, the performer is but the waiter who ideally transfers the abstract truth to the audience.

Performance research however proves that this paper music ideology is erroneous. The score's information is as fragmentary as a culinary recipe without the cook's actualization. The transformational procedure from score symbols to acoustical events is driven by complex continuous parameters of dynamics, articulation, and agogics. Tempo curves and intonation, for example, are essential contributions to a music's semantics see [118, Ch. 36].

16.4 Time Without Now

The European understanding of time is a complex phenomenon that originated in the Medieval construction of omnipresent mechanical clocks, see the history of European time culture [156]. This practice of time perception showed a pointillistic reduction of time: The periodic stroke of the clock stresses the start and end of a time interval, rather than the interval's inner deployment. In the European understanding of time, this reduction lead to a neglect of time as a flowing entity, a time that is only perceived by its endpoints.

As a consequence, the musical concept of presence in the Western tradition has no place except as a zero event: it starts when it ends. The past and the future, conceived in causality and teleology, can be discussed, but the presence as such is a vanishing point in time and thought. The theory of time has no presence, the Now is not conceived.

Recent research [119, Ch. 22.3] however succeeds in designing a conceptual framework of the musical artistic presence. This development also reflects the insights from artists of time, such as musicians, dancers, or actors, that their presence is not a point, but a big imaginary space-time. See Chap. 13 for our discussion of imaginary time.

16.5 Sub Specie Aeternitatis: The Devil of Improvisation

In our Western music schools, we learn a score-based ideology of musical creativity. The divine eternity that wants to be represented in a score's symbols is understood as being the result of a deep meditation, of an ingenious construction following a logical architecture. This concept of a musical composition resembles more the proof of a mathematical theorem than the creation of living art. Its ontology resides in the eternal platonic or Christian heavens, where time is a logical category rather than an existential force. Similar to mathematics, a valid composition can therefore not be created in real time, let alone from the presence of a living musical creator.

As the Italian etymology of "improvisation", namely "all improvviso", "in the way of the unforeseeable", explains, improvisation shares a negative connotation, namely that one is producing from what cannot be planned, thought, or understood. One acts as a random generator. It is evident that this approach is invalid. It lacks all those high level components, which were core to the Christian tradition of thoughtful construction. In a more ethical vein, the improvisor in this perspective is disrupted from all good principles of divine transpiration. It is a devilish practice, bare of intellectual quality and thoughtful planning. We have experienced statements of prominent contemporary composers disqualifying improvisation as a primitive selfsatisfactory activity, as a morally invalid (devilish?) approach to musical creativity. This is all the more remarkable as even Schoenberg conceived composition as a slow motion improvisation.

It is evident, that this bias of improvisation results from a total misconception of this style of creativity. In particular it results from the neglect of the rich space-time of artistic presence as discussed in the previous Sect. 16.4. This presence shares three dimensions (see Fig. 16.1):

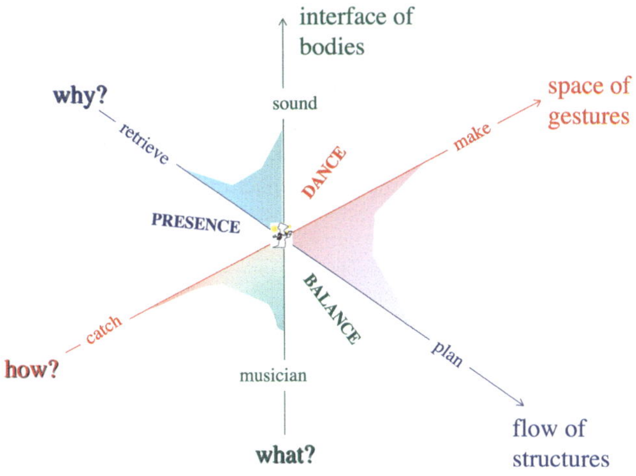

Fig. 16.1 The three dimensions of artistic presence. © Guerino Mazzola

- the flow of structures from retrieval to planning, answering the question “why?”
- the interface of bodies, the instrument between the musician and the sound, answering the question “what?”
- the space of gestures, catching a gesture and continuing, making a new one, answering the question “how?”

The combination of any two of them defines dance, balance, and presence, three essential components of the improvising creator, and everything but random. It would be utterly offensive to qualify John Coltrane, Cecil Taylor, or Miles Davis as being random machines. And it would also be a deep misunderstanding of Beethoven or Mozart who were exquisite improvisors.

16.6 Expert Music Only

In the Western music culture, the musician and the composer have become experts who have to transgress a complicated educational process to achieve their competence. They are distinguished professionals. The average music lover is a passive consumer. In a classical concert, the audience must sit in their chairs and listen to the professionals’ performance. The music media are produced and distributed by professionals. The pop music industry reduces the consumption of music to a little bit of dancing and mainly listening: the very concept of “audience” means just to listen. Music is a professional industry with its highly developed agents: critics, festival organizers, politically supported initiatives, and music networks, see Chap. 8.

Opposed to this tradition, African music culture for example also shares professional musicians, but beyond this, music is shared with everybody, children make music, dance, create lyrics with sound from their early childhood, see [5, 28]. Musical eduction in Ghana typically is initiated as a social activity that does not require an institutional background, it takes place in every family’s environment.

In view of these cultural differences, one should rethink the role of the non-professional music lover, not only in terms of sociology, but also in terms of software and music instrument development, offering PCs not only as personal computers, but also as personal composition machines.

Chapter 17
Improvisation and the Synthesis Project on the Presto Software

17.1 The Role of Improvisation for the Future of Musical Creativity

In [119, Section 22.1], improvisation was analyzed and then defined as a part of an equation

$$composition + improvisation = soundingmusic,$$

where composition was the totality of explicit and implicit rules in the construction of music. This setup however did not deal with the positive characterization of what constitutes improvisation. In [119, Section 22.3], this open question was discussed and answered by the setup of artistic presence, an architecture of the Now, which we also describe in Sect. 16.5.

In view of this perspective of the artistic presence, the role of improvisation for the future of musical creativity should be investigated and developed as a cultivation of the three basic dimensions (flow of structures, interface of bodies, space of gestures) and their combinations (dance, balance, presence). This approach is neutral with respect to musical styles, i.e., it can be applied to jazz, pop, folk music, or classical musics alike.

In the following section we want to discuss a situation, where the three dimensions, flow of structures, interface of bodies, and space of gestures are particularly present and problematic in a positive sense: human improvisation together with computer-generated structures in performance. In this combination, the confrontation of composed structures, which are frozen in a music computer program, with the gestural human interaction of a pianist, presents a challenging force field and also a paradigm of future musical creativity.

G. Mazzola et al., *The Future of Music*,
https://doi.org/10.1007/978-3-030-39709-8_17

17.2 Software Construction and Improvisation: Mazzola's *Synthesis* Project

In this section, we want to discuss the interaction of software-driven composition with human improvisation. We first present the software composition and then the improvisational interaction of a jazz pianist (Mazzola) with this composition. This discussion will shed light onto the future sociology of computer-aided musicianship.

17.2.1 Principles of the Project

This project was funded by a grant from the city of Zurich in 1990. The project's goal was to perform a jazz concert combining live piano with computer-driven music at the International Jazz Festival in Zurich 1991. Mazzola would apply results and methods of mathematical music theory and their implementation in the composition software *presto*®, running on Atari computers. The musical idea was to shape the entire percussive part of the composition via this software. The musical style was planned to be a synthesis of Miles Davis' rock jazz and Cecil Taylor's free jazz, hence the composition's title. The computer-aided construction took 4 months of work, learning to collaborate as a pianist with the computerized composition took 2 months. The recording of the composition for a CD [109] took 1 day, see Fig. 17.1. The composition was performed at the International Jazz Festival Fall 1991.

Fig. 17.1 The cover of the *Synthesis* CD. © Guerino Mazzola

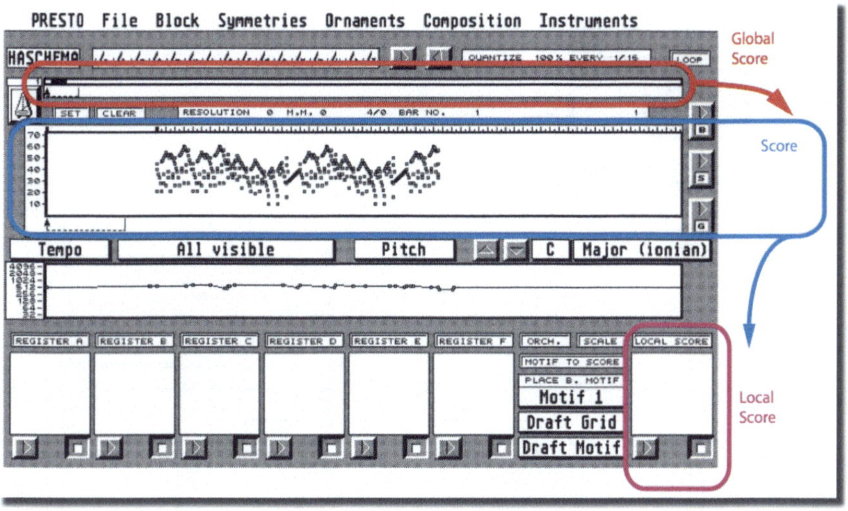

Fig. 17.2 *presto*®'s main display with the global score on top, a finer selection (score) in the middle, and some smaller local score perspectives at the bottom. © Guerino Mazzola

17.2.2 The presto® Software

The *presto*® software was programmed for Atari computers[1] in Ansi C by Mazzola and collaborators 1986–1989 in collaboration with the Fraunhofer Company. The software was programmed to perform some important operations defined in mathematical music theory. It could import and save MIDI files or create music scores by drawing with the mouse or applying geometric transformations and morphing to given sets of notes. Figure 17.2 shows the main display with the global score on top, a finer selection (score) in the middle, and some smaller local score perspectives at the bottom.

The local score can be enlarged to enable graphical interaction (drawing), affine transformations, and instrumental settings on any of the six possible two-dimensional parameter spaces of onset, pitch, duration, and loudness, see Fig. 17.3.

Figure 17.4 show the representation of the *presto*® score, together with some lead sheet type indications for the pianist. © Guerino Mazzola

[1]On an Atari emulation for PCs or Apple computers the program is still operational in 2018.

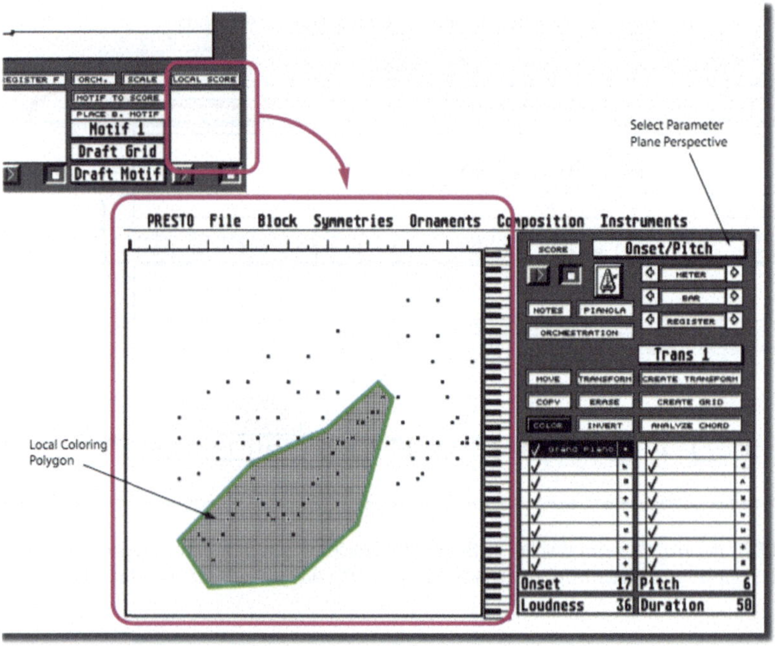

Fig. 17.3 *presto*®'s local score for drawing, transformations, and instrumental settings. © Guerino Mazzola

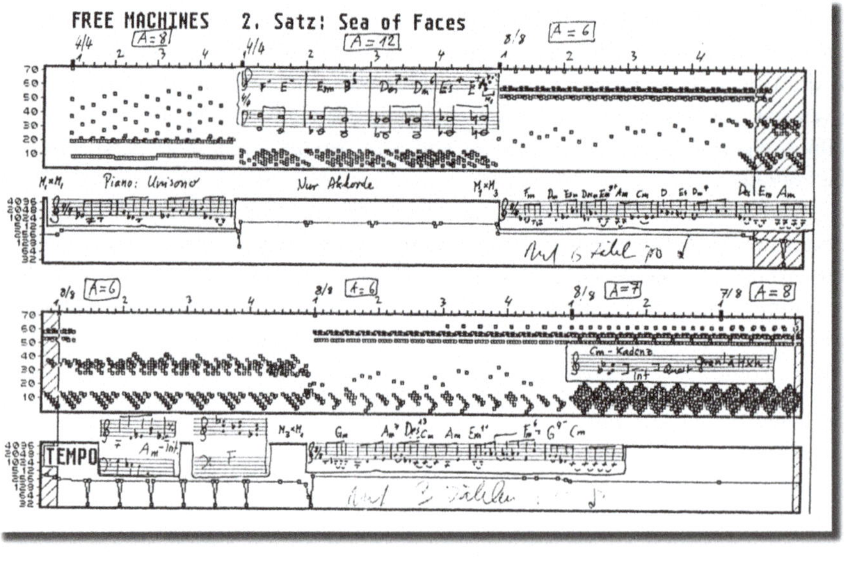

Fig. 17.4 The representation of the *presto*® score, together with some lead sheet type indications for the pianist. © Guerino Mazzola

17.2.3 The Overall Architecture of the Composition

The overall architecture is that of a four movement concert for piano and percussion. Percussion is programmed on *presto*®, the piano is played by jazz pianist Mazzola, see Fig. 17.5. The four movements are generated from the background structure of the 26 three-element motives (classification in mathematical music theory, see [108, Ch. 4.3.2]). From these motive classes, the third movement creates a translation of a poem's letters to a sequence of percussive motives. The other movements also make use of a germinal melody. This melody is a patchwork of representatives of all 26 motive classes, as shown in Fig. 17.6. The first movement essentially realizes rhythmical modulations (in the same spirit as tonal pitch modulations, but rotated by 90° into time) in a sonata form. The second movement is a variational setting taking morphed variations of the germinal melody according to grids of attractive forces defined by Messiaen modes. The fourth movement is a fractal refinement process operating upon the germinal melody. We shall only further elaborate on the second and third movements. For the percussive instrumentation, we could use around 200 instruments from a set of synthesizers. This is of course beyond any

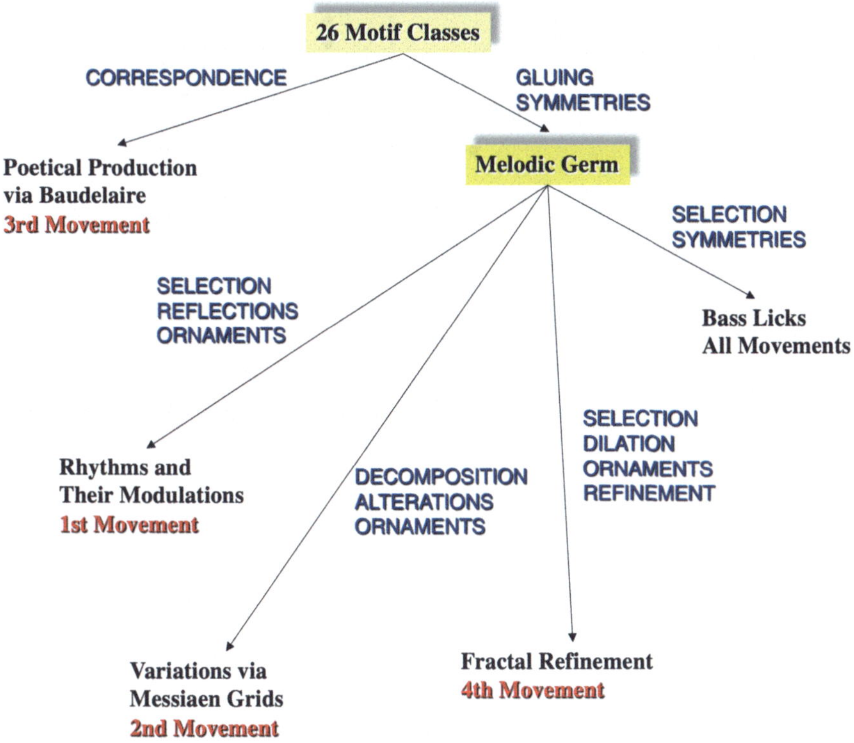

Fig. 17.5 The four movements of *Synthesis*. © Guerino Mazzola

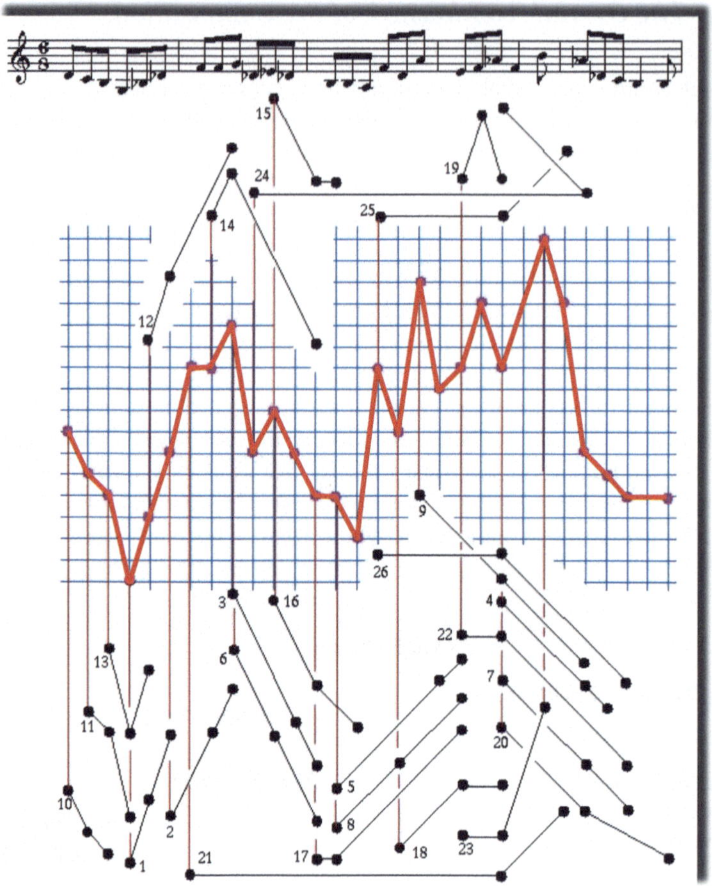

Fig. 17.6 The germinal melody of *Synthesis* is a covering by all 26 three-element motive classes. © Guerino Mazzola

human orchestration, apart from the structural complexity that is realized by the software's algorithms.

17.2.4 Symmetries in Music

It is imperative to recognize the importance of symmetries for musical structures. Some symmetries, such as vertical reflection in pitch inversion, horizontal reflection in retrograde, translation in pitch or time in transposition and da capo, or shearing in arpeggio, are classical examples. But more general symmetries, such as rotations, have also been used in the twentieth century. It can be shown that every symmetry can be written as a composition of a sequence of musically meaningful symmetries (see the concatenation principle in [118, Ch. 8.3]). For this reason, the *Synthesis* composition in its rendition by *presto*® makes extensive use of symmetries and grids that are generated from symmetry operations. This usage greatly exceeds the capacity of human capabilities, reliability, and patience and therefore presents a strong argument for the compositional usage of algorithms and their implementation in music computers.

17.2.5 Second Movement: Morphing Melodies

The second movement is a sequence of structures that are built around morphed versions of the germinal motive, see Fig. 17.7. We first introduce grids that are defined by two dimensions: pitch and onset. In each dimension we place a Messiaen mode, for example (top left example in the figure) vertically Messiaen 1 (whole tone scale) and horizontally Messiaen 2 (complement of the diminished seventh chord). We then rotate this grid by 45° and stretch it by $\sqrt{2}$. This grid is set as a background gravitational force that attracts the melody's notes (connected by red lines) so that a new melody is generated: see top right melody. The same method is also applied to other Messiaen combinations, such as Messiaen 3 vertically and horizontally at the bottom left of the figure. This method generates a set of morphed versions of the original germinal melody. In the second movement, we also play the Messiaen grids as rhythmical structures to represent their gravitational forces.

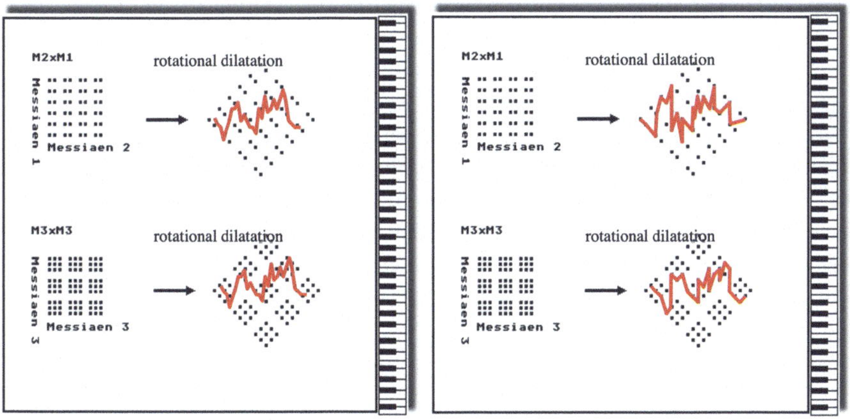

Fig. 17.7 The variations in the second movement are morphed version (red graphics to the right) of the germinal motive, generated by "gravitational attraction" from Messiaen grids. © Guerino Mazzola

17.2.6 Third Movement: The Music of Poetry

The third movement is very different from the others since it sounds like free jazz, in particular, there is no regular rhythm. Its construction starts from Baudelaire's poem *La mort des artistes* from his work *Les fleurs du mal*. We take the collection of letters, which also has 26 elements, and we associate to each of them—following a letter frequency argument and a geometric motive hierarchy argument—a percussive motive from the list of 26 three-element motive classes. This translates the linguistic poem into a rhythmical poem, and we shape the tempo curve of this translation according to the prosody of the spoken poem. This is possible since *presto*® enables a very easy graphical editing of tempo curves.

17.2.7 Improvisation with the Software Construction: Turing's Test, the CD, and Some Critique

For Mazzola, who is a trained jazz pianist and specializes in free jazz, this collaboration with a musically demanding complex rhythmical rendition of a total duration of 45 min was a challenge not only because the machine would not follow any of the pianist's tempo or pitch or dynamic variations, but also because the dialogue was of a rather sophisticated nature: the pianist had composed the percussive part and therefore knew all of its components, hence the dialogue was a kind of memory game to express the pianist's relationship to his own creative utterances. The overall experience was something like flying around one's own ideas and playing with comments and reflections around such fixed points. It was a clear proof that composition and improvisation need not be disjoint musical realities, even in the radical context of compositions being performed by machines.

The production of the *Synthesis* CD was not only the usual precious trace of one's creativity, it was also a Turing test for a sophisticated music software. Recall that the Turing test is about whether a human in a formalized dialogue with a partner can recognize whether that partner is human or a computer. In our case, the CD booklet did not inform the user about the computerized part of this production. The percussion was described as being performed by three musicians, and all "musicians" were represented by a nice drawing by artist Linda Graedel. The general critique was positive about the high quality of the recording because the computer-generated sounds were very easy to trace in high quality on tape. Some critics were however furious that the two styles, Miles Davis and Cecil Taylor, rocking rhythms and free flying sound lines, had been mixed. This proved that style mixes are not always accepted.

But the best Turing test was given when Mazzola performed the piece at the International Zurich Jazz Festival. In that concert you could evidently recognize the pianist and the hardware setup. This generated some funny reactions. After that concert a very qualified jazz critic, who was also a good jazz drummer, approached

Mazzola and told him that the concert was ok, but the CD with the human musicians (!) was way better. This is a remarkable result, not only a positive Turing test, but also showing how important the gestural presence of musicians can be for the judgment of a listener.

Chapter 18
Art Making as Research

Many Black diasporic art aesthetics developed in opposition to colonial projects–emancipation, liberty, humanhood—came about during the Middle Atlantic Passage [39]. These aesthetics cannot be separated from their homelands in Africa, were changed through colonialism and capitalism, and yet still further developed as resistance to dominant systems of power. That is, even after slavery in America, music has continued to be a rich source of inspiration and celebration for African Americans. Here, resistance can be observed in Billy Holiday's "Strange Fruit" during Jim Crow, Sun Ra's "Space is the Place" during the Black Arts Movement, and in Childish Gambino's "This is America" in present-day struggles with social, economic, and political inequalities.

We ask: What is at stake in relegating Black musical performance as solely entertainment and only relatively recently applied to higher modes of inquiry such as mathematics and physics? A scholar cannot trace the significance of Black music in the United States without realizing genres such as jazz, blues, funk, soul, R&B, rock, reggae, Hip Hop, house, etc.–in short, the majority of popular music originates from the African American community. This fact demonstrates a rich history of material to draw on for discovery of new methodologies and application to academic disciplines.

18.1 Choreography, Composition, and Improvisation in Music and Dance

Choreographers who work with composers approach the creation of performance in many ways (Fig. 18.1) such as

1. the composer creates a score that the choreographer follows,
2. the composer and choreographer work in a dialogic process to construct a performance,

G. Mazzola et al., *The Future of Music*,
https://doi.org/10.1007/978-3-030-39709-8_18

Fig. 18.1 "Eat, Prey, Shove"—Choreographer: Jason "J-Sun" Noer for Mixtape Side B at the Cowles Center for Dance and the Performing Arts, 2018. Dancers: Bella Roberts, Averie Mitchell, Iman Sifer-Allah Griffin, Khadijah Sifer-Allah Griffin, Alexander Goode, and J-Sun. Photo credit: Bill Cameron

3. or composer and choreographer create works separately and let a collision of movement and music happen on opening night.

The first approach is the most common but limits the creativity of both artists. When the second method is used as a form of collaboration, the results are closer to a way of understanding how different disciplines can work together. The third is a postmodern approach used by artists like John Cage and Merce Cunningham to create "chance operations" [155].

Improvisation as experimentation is accepted as a mode of inquiry in classical music composition. Regimented scores, defined fundamentals, and mathematical theories first begin with musical experimentation. Black diasporic music rigorously explores improvisation in creation and in an additional way compared to the classical approach which is performance. We can look at jazz lead sheets as an example of always including improvisation as a key element to a composition with definable boundaries.

Improvisation has explicit and implicit rules, can never be exactly duplicated, happens in the moment of playing, and involves decision-making. A musician uses a set of rules to be understood as a specific type of improviser; some of these rules are written out while others are agreed on socially, culturally, and historically. The reason that a jazz improviser can be identified as different from a classical improviser are these sets of rules. An improvisation cannot be captured in a written score, whether it is the exact placement of the hands, a particular venue, or the specific instrument, the moment cannot be exactly duplicated. The problem with exact duplication is argued by Walter Benjamin: "Even the most perfect reproduction of a work of art is lacking in one element: its presence in time and space, its unique existence at the place where it happens to be" [12]. These various factors are part of the engagement of the improviser which means the musician must be present in the moment to take part in the action. Finally, agency, perhaps the soul of the art form, is expressed through a decision-making process as the player navigates their way over the landscape of an improvisation.

"Improvisation is everywhere" [44]. Dance theorist, Michelle Heffner Hayes, points out the continual repetitions of habitual movement hide the improvisational nature of such gestures. Consider how you brush your teeth: How do you hold the toothbrush? How do you hold the toothpaste? How high is your arm? Where is your elbow? What happens when you share your space with another person and must negotiate the range of your motions? The improvisational characteristics of habitual movements can be broken down in performance and examined closely. Such examinations reveal social, cultural, and historical precedents for specific habits and reasons for decisions. Thus, improvisation is everywhere and many habits are not 'natural'. Choreography, composition, and improvisation play crucial roles in the creation of music and dance. It is these aspects that we use to understand how these art forms can be modes of academic research.

18.2 Practice as Research

But can artistic performance be a method of research which yields methodologies and findings applicable to academic areas of study? Research, in this case, is the specifying of a particular issue embedded within an artistic practice which can be examined from many angles by that practice and expanded to other fields. The example of counterpoint see Chap. 15 is a strong argument for the application of a music-based mode of research with clear results.

Dance choreography and improvisation provide another way to consider practice-based research as a valid form of academic investigation. Andre Lepecki argues, "Such understanding of choreography obviously implies that, as with any system of command, choreography also implements, needs, produces, and reproduces whole systems of obedience" [101]. These systems of control are the surveillance devices within phones and personal computers which track our every movement. Choreopolitics are the various ways this movement is restricted and legislated in our society [101, p. 15]. A de-centralized authority is not located in people or actions, "but it is a generalized function of power, an abstract machine holding in place the social order, actually defining the social order as nothing other than a policed thing" [101, p. 19]. Choreopolicing is "defined as any movement incapable of breaking the endless reproduction of an imposed circulation of consensual subjectivity, where to be is to fit a prechoreographed pattern of circulation, corporeality, and belonging"[101, p. 20].

Fig. 18.2 "Pieces of We"—Choreographer: Jason "J-Sun" Noer for Mixtape Side A at the Cowles Center for Dance and the Performing Arts, MN, 2017. Dancers: Charles "Stepchild" Thorstad and J-Sun. Photo Credit: Bill Cameron

That is to say, because a choreographer (see Fig. 18.2) understands how to control a system, they know how to challenge that very system. The dancer in a choreographic work which uses improvisation becomes the visible resistance and exposes the rules of the system. We can apply this approach of choreopolitics and choreopolicing to a wide range of fields which examine how power circulates, maintains, and sustains control over various populations.

We can easily apply this approach to composers as they also employ "systems of obedience" in the creation of a score—scoreography. The musicians translate the score to performance, use decision-making to realize the vision of the composer, and make visible the rules of the system. The musicians use scoreopolicing—improvisation in the audial and physical playing of the music, as defined earlier in this section—to make the audience aware of the boundaries of the score. A specific composition from a genre, such as early New Orleans jazz, can reveal the societal structures, pressures, and imaginaries through performance.

That is to say, it is critical to know the history of classical and African American music and to be able to place New Orleans jazz in this continuum. In doing so, we see the breaking and continuing of tradition, frustration with inequality, and celebration of community located within a specific time, space, and place. Further, how different bodies move in the playing of New Orleans jazz is another indicator of societal rules. Such an intentional positioning provides the context and significance which echoes throughout African American performance in the United States before and after this moment.

18.3 Methodology: An Example and Extension from Dance Studies

Counterpoint was previously mentioned in Chap. 16 as a concrete example of art-as-research, but what are other ways in which composition and improvisation can have a solid impact on the hard sciences? According to dance studies, power dynamics are revealed through close reading and performing of habitualized movements. A rigorous embodied approach allows for new knowledge about these gestures to be produced through a physical exploration of movement. Dance-based research is rooted in a deep knowledge of a dance form which can only be gained through a long history of the practice. A sustained dance practice can be compared to the repetition in everyday movement like walking which, over time, appears as "natural." Such movement is, in fact, the result of thousands of hours of social and cultural immersion. An epistemic vigilance is necessary to question the ways in which we know things. This vigilance not only makes 'natural' behaviors visible, but also opens up questions about what other things we take for granted, like power structures, and how these are created and sustained over time until challenging them seems unimaginable.

Applying a research methodology to a specific dance practice means to use a movement technique (i.e., in break(danc)ing, crossing and uncrossing arms in front of one's body dancing while upright), take it apart step by step (i.e., arms begin to approach crossing from below the waist), think through the implications of changing such movements (what does it mean not to cross arms in dance form focused on battling?), change them (don't cross), and begin theorizing (i.e., what sorts of crossings are happening in social spaces and how does changing these collisions affect the space?). This choreographic process is a blueprint for break(danc)ing as research to investigate movement from a mover's point of view which provides ways to investigate and expand ways of thinking. In this way, a dance-as-research approach can contribute to the broader field of performance studies, as well as, sociology, anthropology, and ethnography among other areas of study which situate movement culturally and socially.

This methodology originates from a specific embodied analysis developed by Jason "J-Sun" Noer (see Fig. 18.3) which rigorously investigates an issue through movement to produce new knowledge. The approach follows three steps:

1. using specific dance techniques to engage with issues like circulations of power,
2. making movement choices which are not part of the accepted canon of technique in order to theorize, and
3. documenting the new knowledge produced from the approach in written, spoken, or bodily form [131].

How can music studies benefit from this dance studies approach? We acknowledge that there are major differences between disciplines but argue the spaces of overlap provide inspiration to develop new strategies for methodological approaches. Another important intersection between many artistic disciplines is

Fig. 18.3 Jason "J-Sun" Noer and Yan Pang—"Current Issues" at Mixtape Part III at The Cowles Center for Dance and the Performing Arts on May 9, 2019. Photo Credit: Bill Cameron

Fig. 18.4 "Everyday Work"—Choreographer: Jason "J-Sun" Noer for Hip Hop? at the Cowles Center for Dance and the Performing Arts, 2012. Dancers: Charles "Stepchild" Thorstad, J-Sun, Anne "AnnieUp" Aldag, and Amirah Sackett. Photo credit: Adam Adolphus

the abundance of academics producing literature about subjects in which they do not take part. The disconnect between academics and practitioners of artistic expressions, such as jazz music or Hip Hop dance, is caused by the fact that the majority of people who write about forms are not doers (see Fig. 18.4). A positionality as a researcher, academic, and long-term practitioner allows for a distinct perspective and use of a practice as a way to understand the circulation of different types of power.

Many non-practitioners produce brilliant and insightful literature which form linkages between these two groups and should not be devalued. However, being a practitioner provides a researcher with an element of experientiality, of embodiment, of an intimacy that cannot be overlooked or understated. After all, how can one theorize about an artistic expression detached from the practitioner's body?

Artistic expressions, particularly Black art forms such as jazz music and Hip Hop dance, have gained visibility in academic study in recent years, but scholarship is lacking in regard to the production and use of such practices as research. Even more scarce is the presence of actual and active practitioners in the fields of music theory, performance, and dance studies. This last point is important to the writers of this book as we are all practitioners and advocate for more experiential research.

Chapter 19
Human and Machine Music

The development of musical production has been to delegate the realization of sound partly or totally to machines. Every musical instrument, except the human voice and body, is in part a collaboration with more or less complex sound generating machines. In what follows we want to discuss three historic examples: Lyra and monochord, steam-powered organ, and panharmonicon. The original Pythagorean music was realized using the lyra and monochord.

In the seventeenth century, universalist Athanasius Kircher designed a technologically demanding steam-powered organ.

The inventor of the metronome, Johann Nepomuk Maelzel, also constructed an orchestra machine, his panharmonicon, and it was Beethoven who composed music, *Wellington's Victory*, for this machine, to celebrate the victory of English troops over the Spanish in 1813.

A recent technological innovation has been presented by China's state-run Xinhua news agency. It has news anchors—virtual presenters that use Artificial Intelligence (AI) to read the news. They are life-like representations whose movements are synthesized from real humans, and their words are constructed from news resources.

The simulation of the human voice has been used for the artificial singer Hatsune Miku, who was constructed for a rhythm-based video game, has over a million followers (!) and has 'composed' more than 10,000 songs using the Yamaha's Vocaloid software [93].

Another artificial being is Lil Miquela, an Instagram model. These technological advancements mean humans and artificial singers will share visual and musical space in the future. It is possible that artificial beings will become a normalized part of global society very soon. To develop this discussion further, a short history of AI development is necessary.

G. Mazzola et al., *The Future of Music*,
https://doi.org/10.1007/978-3-030-39709-8_19

19.1 Artificial Intelligence

Before discussing the fundamental question of why we have begun to delegate more and more of musical production and performance to machines, we have to briefly clarify the hype of AI (Artificial Intelligence). This field of computer science aims at simulating human attributes, especially those which we subsume under "intelligence." AI research started with the simulation of the brain's thinking capacities, such as calculating mathematical formulas. Then, when researchers understood that human intelligence is not the brain's exclusive competence, they extended AI in a second phase of connectionism, where the neural networks were added to the brain's functions. Since this approach was not really successful, the third phase of AI, embodied AI, was created. It added the body as physical expressivity of intelligence to the brain and the nerves, see [139] for a history of AI.

The most prominent AI enhancement of music has been added recently, when Warner Brothers signed a contract with the German computer music company Endel [60] about a software that produces "music" for four different moods: Relax, Focus, On-The-Go, Sleep. The app will play this mood-music for you according to your daily parameters, such as the time of day and the weather. The musical output is simply annoying. There is no audible intelligence in this app.

The present 'success' of AI is based upon the neural network paradigm of connectionism, which is a learning machine approach by training artificial neural networks. These machines never understand anything they 'learn' because AI has no semiotic dimension whatsoever. They only add up big numbers of shapes and may adopt this data to associate a new input with the given repertory. Such a mechanism may be impressive at times, but it does not comply with what "understanding" means: to execute a semiotic activity, going from expression to content, and back. This pattern recognition approach has nothing to do with intelligence in the original sense of the Latin word "inter-legere", to select/read between the lines. What is between must come from an added semantic value, which at present is not implemented in any AI machine. There is not even a mathematical semiotics, let alone a software implementation.

When we focus on AI in music, it is evident that understanding musical expressions, transforming sound into meaningful messages, requires a lot of human intelligence and cannot, at the time being, be delegated to the superficial technology of neural networks. The present neural network hype should also be dealt with in a critical spirit in view of a remarkable theorem by George Cybenko [24]: For any given input, one can construct a neural network that produces any desired output.

19.2 Some AI Components in Music

Apart from mechanical or electronic hardware components of the music production, a number of software-oriented devices have been invented. Kircher, whom we mentioned earlier, already conceived a composition machine, his *arca musarythmica*,

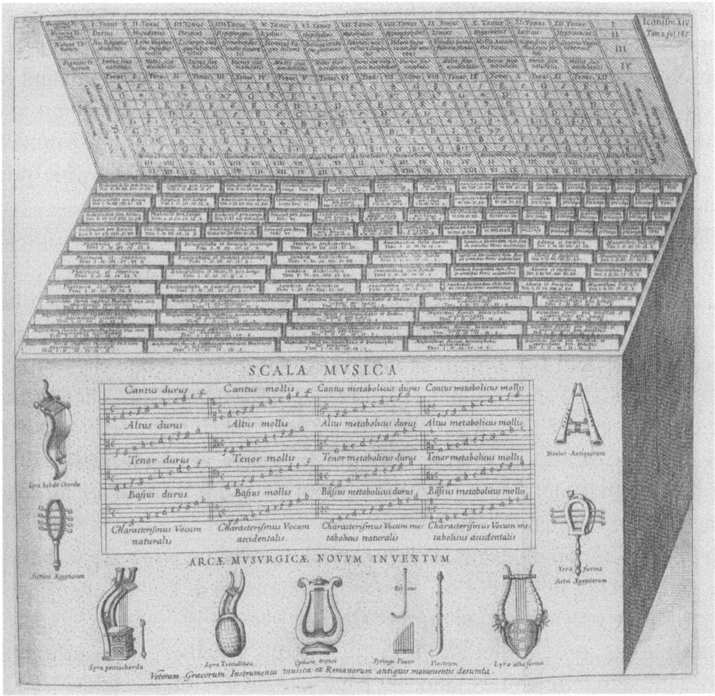

Fig. 19.1 Kircher's arca musarythmica was a device for combinatorial composition. © Herzog August Bibliothek Wolfenbüttel: 1.2 Musica 2° (2), Faltblatt nach S. 185

see Fig. 19.1. The device would work on a huge combinatorial method to compose a large number of musical works from elementary components.

Recall that Mozart also conceived such a—however much simpler—combinatorial device, his famous *Würfelspiel* (musical dice game) which would produce compositions by combining a preset list of measures. In recent times, a number of software programs have been implemented which help compose different aspects of music, from sound synthesis to counterpoint or harmonization and melody or rhythm creation. See Chap. 6 for examples.

19.3 The No/Body Problem

Our discussion in this chapter cannot focus on the superficial fact of delegating musical components to machines, but must specify which part of the process of music production is assigned to machines, and how deeply this delegation operates. This is not only a technical topic explaining the components of a music production flowchart, we also have to understand the fundamental ideas behind such "outsourcing" strategies: Is the human component always intact or are humans being superseded by other agents?

To be clear, it is evident that many musical components might and will be taken over by machines. From composition to improvisation to sound synthesis or instrumental extensions and, yes, also robots who might play instruments with more precision than humans. First examples of robotic musicians were presented by Japanese companies, for Toyota's violinist robot.

The first question that comes to mind is: What kind of music do we aspire/accept when we acknowledge that it might be realized (composed, improvised, played) by machines in important portions? Is every music possibly feasible by machines? Let us recapitulate the musical oniontology (see Chap. 2) to understand the question. We have a triple reality, which should make us think about the critical one: psychological reality. Is this what we could delegate to machines? Then we have the semiotic dimension. As already mentioned in Sect. 19.1, it is unlikely that machines will take over the semiotic strand. The dimension of communication is critical. It might happen that the poiesis and the neutral level are taken over by machines, but the aesthesis is human by definition. Why would we want to replace aesthesis by machines? We humans want to enjoy music, it makes no sense to have robots listening and dancing instead of us. Last: the dimension of embodiment, where the gestural origin of musical production is situated, yes, this could be taken over by machines. But not so if we put all the critical points together! If gestural creativity is loaded with semantic values, and the psychological reality is addressed, such a combination might be impossible to be realized by machines, to put it in simple words: we humans want to live and love, we don't want machines to do it for us.

Nevertheless there is a large amount of musical culture that serves the tendency of some music's machine-driven usurpation. Let us have a look at the tendencies in present-day classical music industry. And let us stress that the patriachic perspective remains omnipresent in many decisions about style and thought backing the industry's output. The product sold by this industry is not music, but a set of patterns that serve the business. To begin with, the repertory is narrowed to the classical canon of works, which have been intensely performed and recorded during the twentieth century. This material has created a pattern of expectations. The amateur of classical music expects to hear what has been played in concerts and, more so, on LPs, CDs, cassettes, etc. Aesthesis in this community has become a pattern recognition, not a musical listening. They don't listen to the music, but to the patterns they have memorized. The psychological reality of the audience has been reduced to a recapitulation of what one already experienced, again a pattern recognition, but this time one of emotional stereotypes. They claim to understand because they recognize. This industrially conceived construction importantly includes the staging of musicians in concerts. They have to fulfill the pattern which the empty recognition machinery feeds.

Let us look at an excellent example of this machinery: Pianist Yuja Wang, artist of the year 2017 by the journal Musical America, Pianist Lang Lang or violinist André Rieu would be other examples. To begin with, she is one of the most outstanding performers when it comes to precision, expressivity[1] and velocity. She interprets works of the classical repertoire exactly as the pattern matching machinery requires—impeccable. The staging of this performative professionality needs some comments. The audience knows all those works, here is a dramatic emptyness of such a machinery: they celebrate the infinite repetition of patterns, so why could this be attractive, what are they attracted to, besides the reconfirmation of biases and individual stability? This virtually empty content must be dressed in an optimal style. It is the opposite of the naked king: It is dresses only, the king is absent, well, reduced to a pattern. And Wang is the perfect embodiment of this staging. She is not only a fantastic virtuosa, she is smart, beautiful and intentional. She celebrates her appearance and intentionality explicitly by her style of dressing. According to her comments for an interview in the fashion journal ELLE in June 2016, answering the question about her choice of "fantastic dresses", she states: "It's hard to find clothes because I'm so petite. In my twenties, I'd put on my tight Hervé Léger dress and heels, and it looked like I was going to the bar. Concertgoers think, Classical music—it's really serious. There are lots of rules, and the dress code, which I broke, was one of them. It's irrelevant to what we're doing. It's just a piece of cloth, but once it's on my body, it boosts my confidence, and that translates to the music."

Summarizing, this industrial construction of music business does not sell music, but a well-designed machine of pattern recognition, and on this basis can survive economically. This machinery is exactly what AI can cope with: neural networks are pattern matching devices. This type of musical culture is for machines by machines, even if they look like humans.

[1] Especially in percussive works, not so strong in meditative adagios, her father is a percussionist.

Chapter 20
The Role of Music in the Diversifying Cultures (Africa, East Asia, South Asia)

20.1 Africa: Ghana

This section stresses the rhythmic complexity of Ghanaian music (as representative of African music cultures [5, 133]), which is not only present in the drumming tradition, but also comprises poetry and dance, see Fig. 20.1 for the model as proposed by Kofi Agawu in [4].

This musical expressivity is not restricted to "professionals", but is shared by virtually everybody, from early childhood to adult, see Fig. 20.2. This omnipresence of music is beautifully illustrated by Rwanda's Sherrie Silver, who dances with many children in different African and also non-African countries. See [150] for videos.

The presence of music in Ghana is not only sustained by a distinguished professional artistic population, but shared by everybody. It is a way of existence like breathing, without music you cannot survive. Interestingly, the large University of the city of Kumasi (Kwame Nkrumah University of Science and Technology, KNUST) does not have a music department. KNUST is not only for technology, it has department of Economics, English, Geography, History and Political Studies, Modern Languages, Religious Studies, and Sociology. One might guess that music is so elementary that academic perspectives are not urgent. We could not find out from local authorities why this is the case.

G. Mazzola et al., *The Future of Music*,
https://doi.org/10.1007/978-3-030-39709-8_20

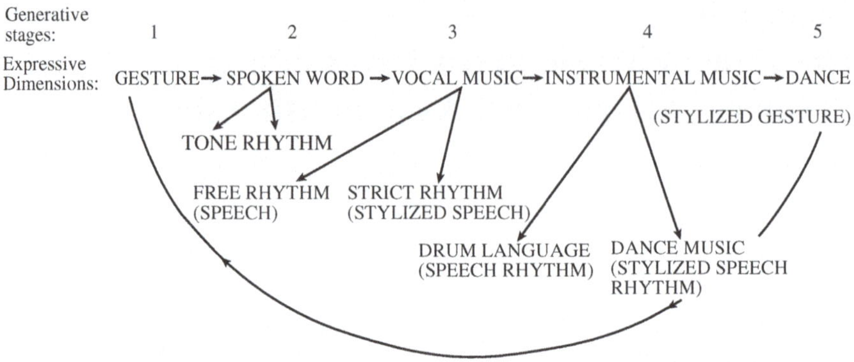

Fig. 20.1 The African rhythm model as proposed by Agawu. Image reprinted with kind permission from Kofi Agawu

Fig. 20.2 African dancers from Ghana. Courtesy of the Penn Museum

20.2 East Asia: China, Japan

The Japanese Noh theater (see also Sect. 21.2.2) is a good starting point to discuss the role of music in East Asian cultures. The Noh theater was invented in China and later imported to Japan and developed as the "Art of Ma" [98]. *Ma* in China originally meant a spatial dimension, but in Japanese Noh theater it was united with a temporal dimension and can be translated by multiple concepts, such as space, spacing, interval, gap, blank, room, pause, rest, time, timing, or opening. An architect may mean space, while a musician intends time. The word for *ma* represents the sun in the middle of a gate.

Ma is however different from Western concepts in that it addresses the inside of space/time, not their boundaries. While in traditional Western musical time concepts, the start and end of a sound are essential, but not the inside of the event, *ma* is the opposite: being inside is the focus. This is evident when a Noh play is unfolding: the duration of a specific event is not focusing on its start and end, but delves into the interior, duration as "end minus start" is irrelevant.

The function of music might be understood from the *ma* concept. In China and Japan, rhythm in the sense of an organization of time as a succession of determined events is secondary. Being inside time, not starting and ending time is the existential position. The function of such an art of time, of *ma*, defines a radically different musical culture from Western or African approaches. Music expresses a meditative existence, a movement inside one's body, not of the body's gestural surface.

20.3 South Asia: Indonesia

This musical culture is very diversified as distributed over one thousand islands and producing hundreds of local styles for a gamelan orchestra with dancers. The history of Indonesian music is also influenced by other cultures, Indian, Chinese, for example. Here we should describe this multiverse in short but adequate thoughts.

Mazzola's experience with Indonesian music stems from the documentary *Teak Leaves at the Temples*, where his free jazz trio (with Heinz Geisser on percussion and Sirone on bass) collaborated with many Indonesian groups from local communities to contemporary popular groups [132], see Fig. 20.3, where Mazzola plays with a local rhythm group in a village that was destroyed by an earthquake.

From that experience, and also with regard to a collaborative project with Indonesian dancer and professor at the University of Minnesota, Diyah Larasati, it

Fig. 20.3 Mazzola playing with a local Indonesian group. © Guerino Mazzola

can be said that Indonesian music is very rich and shared by a dense community of musicians and dancers. Music has also been politically profiled during the military coup in the sixties of the twentieth century. It could be understood as a culture in between the Western expert musicians and the African everybody musicians cultures. In the mentioned documentary, it becomes evident that the gestural component is very important in Indonesian music: human gestures shape time, and that the gestural 'language' can be used to overcome a "clash of civilizations".

Chapter 21
Cultural Theories of Gesture

We approach and define gesture as physical, symbolic, or psychological movement. Physical gesture is the visible manifestation, however minute, of movement originating first with a movement of thought. Symbolic gesture refers to the musical expression that begins with the score but, again, is first recognized in the mind. Psychological gesture is the process of receiving information and the following mental reactions which result from the interaction.

Human gestures are culturally, socially, and historically located, and are therefore specific to location, people, and time. Some gestures convey a relatively common understanding, i.e., a nod usually connotes agreement or affirmation but differ in expression. Comparatively, there are gestures, like the Indian head bobble, that are specific to a culture and do not exist in the same way in other places.

21.1 The Origin of Gesture

There has long been a debate about the origin of gestures in relation to spoken language which remains unresolved. One one side, proponents argue that development of oral communication was developed from physical gestures. Merleau-Ponty's philosophical claims are supported by recent results from cognitive science and neuro-science. Maurizio Gentilucci and Michael C. Corballis have proposed a theory of gradual transition from manual gesture to speech [106]: "It is supposed that speech itself a gestural system rather than an acoustic system, an idea captured by the motor theory of speech perception and articulatory phonology. Studies of primate premotor cortex, and, in particular, of the so-called 'mirror system' suggest a double hand/mouth command system that may have evolved initially in the context of ingestion, and later formed a platform for combined manual and vocal communication. In humans, speech is typically accompanied by manual gesture, speech production itself is influenced by executing or observing hand movements,

G. Mazzola et al., *The Future of Music*,
https://doi.org/10.1007/978-3-030-39709-8_21

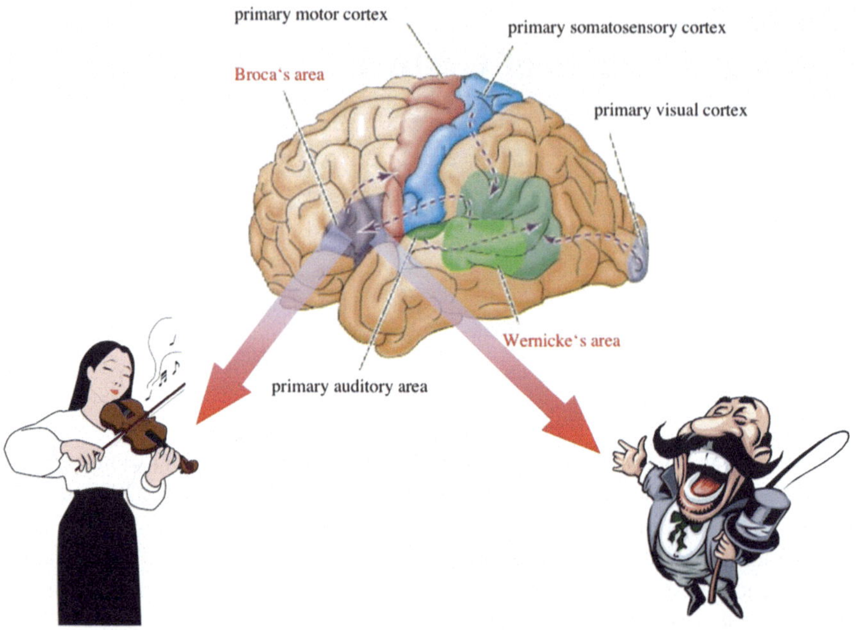

Fig. 21.1 The left hemisphere of the human brain, showing Broca's area and its significance for fine-grained and fast motor control in musicians and the production of speech. © Guerino Mazzola

and manual actions also play an important role in the development of speech, from the babbling stage onwards." The premotor cortex, lies above Broca's area, which is responsible for speech production, as opposed to Wernicke's area, which takes care of speech perception, see Fig. 21.1.

On the other side of the debate, the dispute is rooted in the fact that gestures must be in a line of sight and cannot be seen at night. A more generative stance is to understand exclusivity between theories forecloses a flexibility in thought-movement which often gives a more realistic picture of, in this case, the meaning of gestures. Another way to think about the contextuality of gestures is to simply acknowledge movement has accompanied speech for so long in human history that debating the specific origin is, at most, a theoretical exercise. We are interested in the real applications and implication of movement taking place in the interactions between score, musician, instrument, and audience.

Movement is always accompanied by gesture and an intentional viewing is necessary to see these articulations. To an observer, running is a general type of movement if only examining the legs of a mover and, in this example, we must consider the entire body. How are the hands held? Fists or straightened fingers? How do the arms swing? Straight or loosely? Eyes? Head? Shoulders? The previous questions highlight how gesture can be seen in a moving body as an integrated and complex process.

In relation to music, gesture accompanies the musician in the context of every form of music. A performer must move when playing and singing, so the gesture is the, mostly, unacknowledged accompanist in the development of musical theory to the present. A point should be made here about the playing of instruments by performers with disabilities and that gesture is also eye movement and minute muscle control. This type of movement is incorporated into Japanese Noh theater and Butoh dance which uses small physical motions over long periods of time. The frozen expressions on Noh masks are considered, by practitioners, to be extremely expressive; a bowed head can be contemplation of a pond and a lifted head can be looking for the moon. In this example of Noh theater, we can think of the mask as the instrument and the body as a way to contextualize the gesture.

21.2 Gestures in Relation to Culture

21.2.1 Gesture in American Hip Hop DJing

Let us look at different types of music and characterizing gestures that accompany each. American Hip Hop DJs use many movements to connote rhythmic awareness which is the unifying element in this form of music. Q-Bert is a renown turntablist, innovator of techniques, and professor at the Skratch University. He is a prime example of gestural expressiveness within a form almost exclusively performed standing which allows for the entire body to move.

Consider his 1991 second-place set for the world renowned Disco Mixer Competition (DMC) [27]: Fingers pull and push vinyl records using precise amounts of pressure to elicit a long or short distortion, known as a scratch, and accompanied by horizontal movement of a mixer to switch back and forth between turntables. Mind-bending speed blurs his hands and his head angles to right each time the transformer scratch is performed. Q-Bert uses gestures like rhythmic head nods, exaggerated arm movements to accent scratches, and purposeful glances at the audience to emphasize sonic manipulations. Critical theorist, Fred Moten, has said, “Your ass is in what you sing” [128] and Q-Bert demonstrates a complexity of asses with different samples combining in new arrangements.

21.2.2 The Sign Language Formalism of the Noh Theater

The classic Japanese Noh theater is originated in China with the Nuo theater during the Tang dynasty (617–907) and was developed in Japan by Kan’ami and his son Zeami Motokiyo in the fourteenth century. The gestural principles of the Noh theater puts in evidence the fact that the gestures are not necessarily semiotic vehicles in a classical art. This description will then allow to pass to a justification of our definition of the gesture that will follow.

The Noh is made with many different elements and it is an art composed of various media, each of which plays an equilibrated role as a whole. It is necessary to discern the three types of human expression: the expression of the voice, the face and the gestural. Since actors wear masks the facial is impossible. The voice is ambiguous in Japanese, so it is inappropriate for the expression of will and therefore reserved to lyrical evocations. Only the gestures remain then. Therefore Noh is a dance drama, and not an opera or ballet. Its overall shape is extremely rigid and architectural.

In this spirit of rigid forms, the gestures of Noh are reduced to the unit kata and symbolic renditions. This allows a richer communication with the public than through daily actions. Here is the classification of the kata, see Fig. 21.2.

There are three types of movement units:

- realistic, e.g. holding a book in your hands, jumping into a bell
- symbolic, e.g. crying = holding your hand to your face and removing it (see Fig. 21.3)
- abstract, e.g. zigzag = start and the end of many dances

What is important for us is the characterisation of the gestures of human (Shite) and the villain (Waki): the human weaves a fantasy texture through curves. The villain draws the reality with straight lines. This gives us a first approximation of the mathematical concept of a gesture: we will consider curves, whose shape (sinuous or straight) describes the two main characters. It is remarkable that an art as old as Noh is aware of the mathematical form associated with the concept of gesture.

The Noh does not follow a Western pattern of semiotics: the semantic is reversed here. The gestures are used to show the expression of movements inside of space-

Fig. 21.2 *Kata*: Forms of movements in Noh. Image reprinted with kind permission from Delatour France

Fig. 21.3 Crying. Image reprinted with kind permission from Delatour France

time, the ordinary semantics is of second order. The movement of gestures in this space-time is a system of curves that are used for the exploration of this space-time and not for ordinary statements. The story only serves as a vehicle for this set of forms.

In Noh gestures, it is remarkable that the narrative is just the formal carrier for gestures as movements, semiotics are reversed: stories for gestural forms, not forms for stories. There are also gestures not directly tied to sound production but that help orchestral musicians select the best specific gestures to bring forth a general musical idea. This is the role of an orchestral conductor. We define symbolic gestures the movements suggested by a score that ideally allow a perfect realization of the content of score. We label physical gestures the real movements by performers.

21.2.3 Chinese Gestural Notation and Opera Performance

Chinese language is based on gestures and visual symbols as well as the ancient gestural notation. Western music notation represents the sound (result), but old Chinese music notations represents gestural instructions (production) for the player. The earliest music notation in China was created for an instrument called Guqin (古琴). In addition, because Chinese instruments were created for solo performance, there is no expectation or requirement for precise beats. The analogy might be this: In Taiji (太极), individuals move in response to their own inner mood; however, in Western dance, the individuals respond to the rhythmic pattern from the music outside of their own bodies. This makes the gestural message so difficult to grasp from modern Western notation.

The Sichuan Opera's long history teaches the listener not only about music but also about Chinese art, society, history, and people. In recent times, many modern music fans have forgotten about the Sichuan Opera. Few contemporary compositions are written in the style of traditional Sichuan opera. In addition, not many people outside of China know about the Sichuan Opera. These styles convey the emotional and dramatic content of performances, see Figs. 21.4 and 21.5. The music plays a central role in the performance of Chinese Opera. All stage performances involve some form of singing and musical instruments. The principles of music in drama in Chinese opera will require a lot of physicality of the performers. It requires the efforts of a singer-actor who is a practitioner of Chinese traditional singing techniques and knowledgeable enough to tell a story, the performance stage singing, and dancing. Daily oral conversation is usually considered to be too realistic, thus the dancing and singing elevate the drama. Sichuan Opera is one of the most influential forms of Chinese local opera. The tunes have various distinctive characteristics with rich local style. It mixes aspects of various Chinese musical traditions including Gaoqiang (High-pitched Tunes), Huqin (Urheen), Dengdiao (Light Opera), Tanxi (Tan Opera), Kunqu (Kun Opera), all of which have contributed to the development of Sichuan Opera music [135].

Fig. 21.4 Qiong Liu, Sichuan Opera Performer (Chinese National Performer). Image reprinted with kind permission from Qiong Liu

Fig. 21.5 Qiong Liu, Sichuan Opera Performer (Chinese National Performer). Image reprinted with kind permission from Qiong Liu

21.3 The Role of Gestures in Future Music

There are also gestures not directly tied to sound production but that help orchestral musicians select the best specific gestures to bring forth a general musical idea. For example, the conductor plays a role as a score interpreter, translating the symbolic gesture to a physical one. See also Chap. 14 for the mathematical theory of gestures. We call symbolic gestures the movements indicated in by a score that ideally allow a perfect realization of the content of score. The conductor reads the score and creates physical movements to unite the orchestra. We call physical gestures the real movements by performers and conductors. There are many gestures that travel across cultures and are transformed by people in specific geographic regions. The gesture of strumming/plucking/pressing a button can be traced to different influences and the impacts of artists from different places. The intersection of culture, movement, and art are revealed in the practice of being a musician.

What is the future of gesture in music context? How do we integrate gestures in music? Music is not only about sound. It is about gestures which are related to sound. How can music integrate gestures in performance/composition/improvisation? Gestures have been the origin of music notation (neumes), and many important composers have always created their works from a gestural input on their instruments. In this sense, gestures are the origin of composition and improvisation. Beethoven and Mozart would be good examples. Gestures in performance are essential to transform the abstract symbols of the score to sounding music.

Part V
Creative Strategies

Chapter 22
Recapitulation of Creativity Theory

22.1 Defining Creativity

Creativity has been explored by many scholars. We seek to define the process of creativity as something that can be universally attained. When we create there is a general process which we propose is universally followed by all artists. Notably, this process is reversed for consumers and critics of art.

Artists start with gestures. The stroke of a pen or paintbrush, a pirouette, the wiggling of fingers... All these are gestures. They're not primitive, nor are they mindless.

Gestures are then fit into some sort of process. Perhaps, a pianist chooses to follow a standard blues chord progression. Perhaps, they run a broken rake across their keyboard. Whether or not an artist is anti-establishment makes no difference. The fact is that when there are patterns that can exist, there are also instances in which there seem to be no patterns. Both are possible cases which the artist chooses to do over the span of a series of gestures.

By the end of it all, we reach facts. This level of factual information is comparable to the neutral niveau of Molino's tripartition [130]. It is from this point that one may begin tracing and trying to understand the creative process. An interpreter will only experience the facts (Fig. 22.1). It is their mission to try to understand the processes and gestures and how they constitute the final product and message.

To be creative is a question of its own right. We propose that there are seven types of purposes behind the act of creating something. Typically, they are followed in this order. Examples are given below each step.

1. *Exhibit an open question.*
   What is life? Why is this minority group being repressed? How do I feel?

G. Mazzola et al., *The Future of Music*,
https://doi.org/10.1007/978-3-030-39709-8_22

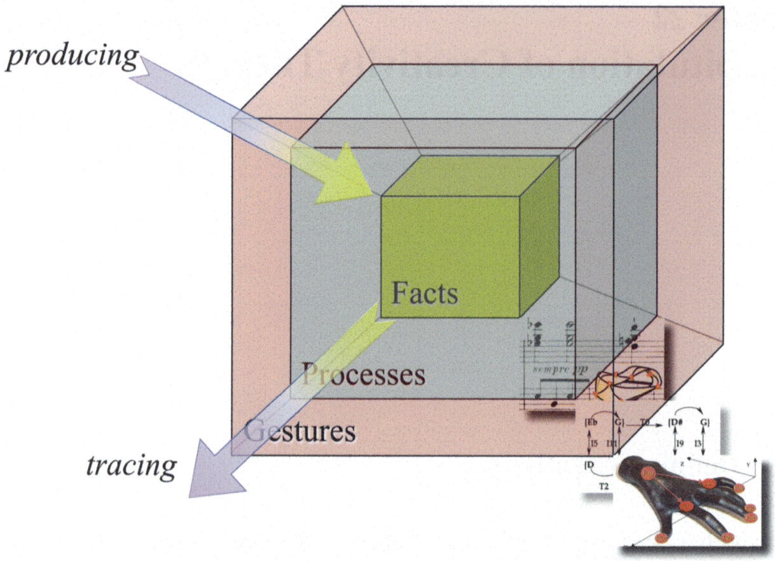

Fig. 22.1 A diagram that outlines the creative process and its reverse. © Guerino Mazzola

2. *Define its context.*
   Children, adults, clones. Africa, Europe, Asia. Antiquity, Renaissance, post-World War II. Where are we and who is around us?
3. *Find a critical concept.*
   Morals, values, social ideals.
4. *Exhibit its "walls".*
   What are the social constructs which limit behavior? How is philosophy and reasoning inhibiting creativity? How could creativity destroy civilization?
5. *Soften the walls.*
   Be the devil's advocate. Provide alternatives. Propose the opposite.
6. *Extend them.*
   Make a new reality. Suppose that the opposite is true.
7. *Evaluate the extended concept.*
   Is this right? Is this wrong? Should we change or adapt to this new reality? Is it even possible?

Chapter 23
The Specifically Musical Walls Against Creativity

23.1 Tradition

Musical tradition is strongly tied to a system with its history and rules. The serial system is an example we discussed in Chap. 11, the system of the first Viennese school comprising Joseph Haydn, Wolfgang Amadeus Mozart, and Ludwig van Beethoven with its contrapuntal, harmonic and motivic specificities is another. See [119, Ch. 13.2] for details.

From the perspective of creativity, the open question (1) is: "What can be the role of a musical system for the composition of valid musical works?"

The context (2) of this question has the following ingredients:

1. They are all built on a set of available sound events, be they defined by pitch collections (such as scales in tonal music) or rhythms or melodic repertories (such as series in dodecaphonism).
2. This material can be manipulated following a set of rules (such as the selection of chords in tonal music or transformation of series in dodecaphonism).
3. These rules must be built according to a defined logic and be logically consistent, i.e. contradictory rule conflicts should not be possible. For example, the contrapuntal rules must be defined in such a way that the composer is never led to dead ends in the contrapuntal construction, i.e. to situations where the continuation of a contrapuntal construction is impossible because of too strong constraints.
4. The rules must be of a certain simplicity to enable constructions of compositions with a reasonable effort. Simplicity also means transparency. A rule system that can only be controlled by a computer is not what we aim at; there is not substantial delegation of musical creation to external devices and authorities.
5. These rules must be communicative in the sense that compositions that follow these rules make them evident to the audience. Rules that are beyond communi-

G. Mazzola et al., *The Future of Music*,
https://doi.org/10.1007/978-3-030-39709-8_23

cation are possibly interesting to the composer in his/her intimate work, but the system has to solve the problem of transferring them into the musical utterance.

6. These rules must be flexible enough that musical creation is possible and even boosted. They must be catalyzers of musical invention and not only directives that kill the composer's fantasy. It is like with traffic rules: They enable maximal freedom within the logic of a working traffic system.
7. The system must help create musical meaning, not just forms. This is in accordance with our principle of creativity as a generator for new meaning in music.

The critical concept (3) here is the "system of rules" within the given context. The system's rules play a mysterious role. They enable what the composer wants to construct, but they also direct and restrict the composer's freedom of choice.

What the critical concept's walls (4)? In such an approach, the subject which applies the rules has been limited. The rules have become formulas that yield output upon any materially admissible input. This is the approach used by American composer Tom Johnson, who has devoted much of his compositional efforts to get rid of the individual and subjective layer of composition. His string quartet *Formula* is typical for this approach: Each of the eight movements is based upon a mathematical formula.

The wall of the rule concept is exactly this seemingly unavoidable constraint: *to apply* the rules. The wall properties of this situation are characteristic for walls: They are so strong and omnipresent that even the question about their necessity and function seems absurd. They are part of the definition of the concept. A rule is there to be applied, what else? But this is exactly the moment where we have to step back and look at the complete context of musical composition. To reduce compositional activity to apply a rule system is a huge step. What happens in this process? Are we playing an empty game, or is the system so strong and intelligent that we are dispensed from rethinking everything once we have understood the logic and power of these rules?

How can we soften these walls (5)? Let us look at a very classical formula: the cadence sequence $T_X \to S_X \to D_X \to T_X$ of tonic, subdominant, and dominant chord functions in a given tonality X. If the essence of the cadence formula were this sequence, any realization thereof by a set of chords would do. But this does not guarantee any interesting musical output. One could argue that voice-leading rules must be added to the cadential formula, and so forth. But is that a sufficient rule set for interesting music? It is not. We are always facing this situation: The set of rules is exhausted, and we still do not have interesting music.

To be clear, we neither defend nor attack the claim that a rule system can produce intelligent music. It is something different that fails. And this is hidden in the meaning of the verb "to apply," which we identified as the wall of the rule system. Can we go to the rule system in its early state where it was, like the neumes, not a dignified authority, but a living organism, so to speak? If "playing" could be resolved by "dancing on the instrumental stage," then the rule system's execution resembles the application of the theater's screenplay. If we may dance the composition, should we then also choreograph it instead of applying the screenplay's prescriptions?

Let us try to extend the wall (6) of rule application with the last section's question on choreography or dance composition. What does a rule system look like if we choreograph it (with autonomy) instead of applying it faithfully? We have to *appropriate* the system; it must become ours instead of us being its slaves or delegates. This means, in particular, that its components must also be given a meaning that we cover with full responsibility. We (the composers) cannot refer to contents that are charged to others—all of it must be ours.

Such an extension must be tested (7), there is no guarantee that an extension as such is a priori successful.

23.2 Extramusical References

It is a very seductive force for composers to create their works as an expression of extramusical contents. A religious composer, such as Johann Sebastian Bach, wrote music to worship God's glory. A romantic synesthetic composer, such as Alexander Scriabin, composed music as an expression of visual phantasies, color symphonies, for example, in Scriabin's case. Or a politically engaged composer might write music to express a political ideology, such as Hanns Eislers, who composed the national anthem of the German Democratic Republic. Or Archie Shepp, the free jazz saxophonist who composed music for the human rights of African Americans.

This fact is however dangerous since the wall here is the hidden implication that the content of the extramusical message should shape a rich musical structure. In the above examples, this was not a limiting or destructive condition because these composers are high ranked musical thinkers. But if a composer believes that the extramusical message is sufficient to guarantee a good musical structure, the enterprise might fail. There are two problems with such an approach. First, it is not clear how the content of the extramusical message should be reflected within the musical work. Second, the structure of the extramusical message may not be transferrable to the musical structure. For example, if we have a linguistic text, how should its syntax be transferred to a musical syntax? Or if the text describes a love story, how would one incorporate the lovers' roles within the musical components? This problem is also virulent when a musical work is interpreted (without referring to the composer's explicit intentions) in terms of an extramusical narrative, replacing the musical (mis)understanding by a fictitious human drama (as done by Schoenberg when he introduced erotic relationships among chords in his harmony book).

The point here is that extramusical contents are by no means a guarantee for good musical compositions, exactly because musical contents are fundamentally different. To recognize such a wall might be hard, but the confusion it creates can be huge.

23.3 Missing Ontological Connections

This is a subtle type of walls. It appears when one does not take care of the specific ontological positions of one's ideas. For example, typical in serialism, one has a poietic construction of highly symbolic value, such as a mathematical transformation between different objects. One composes some music using this construction and does not realize that this construction is not per se musically present. A transformation, for example, might be a mathematical poietic entity, but in the composition, its presence is all but tangible. Here the wall is the failure of switching faithfully between different ontological positions.

Chapter 24
Examples of Creative Extensions in Music

24.1 New Counterpoint

The mathematical theory of counterpoint (see Sect. 15.1) was primarily concerned with the modeling of classical Fuxian counterpoint. The mathematical setup however entailed a more general conceptual framework. A characteristic of mathematical modeling is working with concepts that are genuinely generalizable. The solution of the Fux counterpoint was therefore not mathematically complete, its very concepts enabled a more general modeling context. This change of perspective is crucial here. One changes from the musical perspective (Fux theory) to a mathematical perspective (symmetries, deformations of interval sets, etc.) and *there* rethinks the musical potential, and everything is being done in the mathematical ontology. The transfer of this work to music, generating five new "counterpoint worlds" (see Chap. 15), is the result of such an ontological switch.

24.2 Bitches Brew

Miles Davis created his revolutionary album *Bitches Brew* in 1969 [25] by, among various innovations, a complete reshaping of the instrumental setup. Chick Corea, one of his keyboarders in this music, when showing up in studio, did not find the piano, and asked Miles about where the piano had gone. Miles pointed to an electric piano and commanded Chick to play that instrument which he had to do despite his protests. Miles was fed up with the traditional jazz instruments and reinvented jazz, now known as rock jazz, by a strong step out of the instrumental wall of jazz tradition. With *Bitches Brew*, the jazz-rock message changed the sound of jazz forever.

G. Mazzola et al., *The Future of Music*,
https://doi.org/10.1007/978-3-030-39709-8_24

24.3 Free Jazz

Free jazz was introduced around 1960 by Ornette Coleman, Cecil Taylor, and later also by John Coltrane, Archie Shepp, and others. Why was this movement a creative step? Ornette Coleman named it in the liner notes of his famous album *Free Jazz* [116], writing: "Let's play the music, not the background." He was not only referring to the standard jazz situation, where a group would play in a bar the sonic background for guests consuming their drinks and having the usual smalltalk. He was rather intending to question the role of the given lead sheets for his music. Jazz musicians had to play/improvise following the structural guidelines: jazz scores, blues scheme, song scheme, etc. For Coleman, this was the service of a musical robot or waiter, bringing the music to the guests, but not playing the music, not 'cooking' it in a creative way. He claimed a return to the creative production of music, to the full responsibility for what is being played, not its commercial and institutionalized reproduction (within the narrow limits of prescribed lead sheets).

This was a political statement, but it was much more than a movement of revolutionary African Americans, it was a revolution against the entire system of score-enslaved Western musical culture.

Chapter 25
Performance and Composition

25.1 Performance of Composition

Performing a composition, typically a score in classical music, or a lead sheet in jazz, should be a creative action, not a waiter's role of bringing a dish from Mozart to the audience. This creative difference that defines a valid performance from a mechanical reproduction has been discussed in numerous publications, such as [1] (English translation: [3]), [97], or [111]. They all agree on performance as being a kind of "thawing the frozen gestures in the score." This transformational process can and should be highly creative since it adds to the written template the finesses of performance parameters which cannot be written down, such as agogics (tempo curve of timing microstructure), dynamics and articulation of single notes. Figure 25.1 shows the tempo curve of Vladimir Horowitz's 1987 performance of Schumann's *Träumerei* in beats per minute.

By these performance shaping actions the performing artist adds meaning to the score. This can be done in three different ways: analyzing the composition, interpreting its emotional content, or expressing its gestural components. Performance research has successfully modeled analytical performance. It was possible to simulate performances of famous musicians by applying harmonic, rhythmic and melodic analyses to the shaping of performance, see [111, Ch. 22] for details. It is however not clear, how a composer could add explicit gestural directives to the score. Neumes, the original shapes of our Western notation, were in fact gestural signs, but this aspect has been effaced in the present notation.

What could be the role of the composer in this enterprise? Should composers care about the performances of their compositions? Of course, it is composers' personal decisions to listen to the performances of their compositions, but it should matter if the composers feel responsible about the communication of their works to an interested audience. And, in this case, the manifold of performances can be influenced by composers.

G. Mazzola et al., *The Future of Music*,
https://doi.org/10.1007/978-3-030-39709-8_25

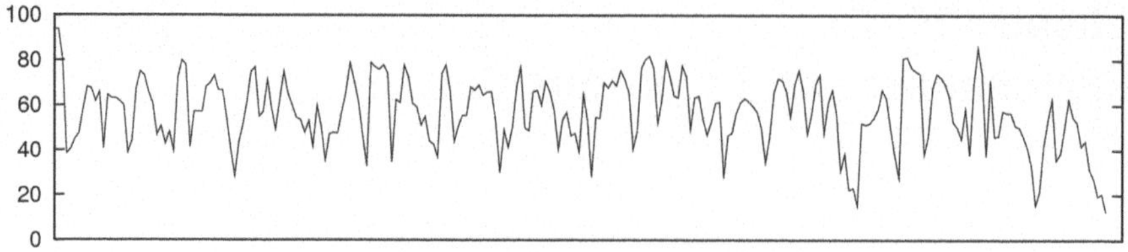

Fig. 25.1 Tempo curve of Vladimir Horowitz's 1987 performance of Schumann's *Träumerei* in beats per minute. © Guerino Mazzola

There is a number of established ways to add performance directives to the score. On the one hand, performance of tempo, dynamics, and articulation can be specified by absolute and relative tempo signs (Maelzel Metronome, accelerando, rallentando, istesso tempo, etc.), absolute and relative dynamic signs (*mf, fff*,..., crescendo, diminuendo, etc.), articulation signs (staccato, legato, sforzato, etc.). On the other hand, extra-musical textual indications, such as *avec une jolie voilée, éclatant, lumineux, etc.* can indicate a performative direction.

Beyond these standard methods, composers should have a competence about what performers of compositions can do with their limbs. Beethoven was a pianist and knew what a pianist of his piano sonatas could achieve. Boulez was not a pianist, and this is felt while trying to perform his *Structures pour deux pianos*. A composer should imagine performative gestures in imaginary time when composing, we have discussed this time topic in Chap. 13.

25.2 Composition of Performance

The less common situation is to think of performance as a compositional enterprise, but there are several cultural perspectives where performance is the moment of compositional creativity, e.g., Indian Raga music or jazz. More generally, what is called improvisation is often paraphrased as "instant composition". And vice versa, composition being understood as "slow motion improvisation". So what is the difference, and can compositional creativity in performance be achieved?

To begin with, instant composition is not what the Italian etymology "all improvviso", out of the unforeseeable, suggests. Improvisers have a huge repertory of musical forms and processes that are invoked without delay. Improvisers are not random generators. They reach a high consciousness of their creativity in the imaginary time while playing. This is where they think about what they are performing and where they make their decisions.

There is, however, a well-known difference, namely that a real-time composition cannot be corrected while playing, as opposed to the compositional standard, where a composer can correct all the points one doesn't like in the score. Some critics of improvisation have argued that because of this difference, improvisation could be of lower quality since errors might subsist. This argument is built upon the idea that music should be judged by some error detection machine, in other words, music should be a construction whose correctness is a decisive quality. But improvisation is not subsumed into this scheme, this has been stressed by Miles Davis, for example. In a conversation with Mazzola, free jazz pianist Cecil Taylor stated: "Jazz is when you walk down the street." This is to say jazz for Taylor is just living one's life, and that cannot be about walking down the street in a correct or wrong way. In this sense, composing performance cannot be equated to performing composition, the compositional style must be radically different.

Are there principles or ideas to be followed when composing performance? To begin with: What would be the composer's perspectives for this type of music? Recalling Mazzola's book on free jazz [116], we may identify three elements: flow, gesture, and spaces. This means that a composer of performance must organize the music's coherence with the flow between the musicians, integrate their gestural communication, and open up spaces of hearing and understanding of what's being played. This is a very different workload from 'offline' composition, it requires an understanding of the musical 'microsociology', the logic of gestural interaction and the difficult culture of listening to *others*. The creativity walls here are built by the attitude stemming from the performance of composition, which presupposes a given material to be followed in a correct way. Evidently, such a wall also extends to a false expectation from the audience, which could be confused when hearing music that is completely self-referential instead of reifying a given material.

From these insights one learns that performance of composition and composition of performance are radically different enterprises of compositional and performative characters.

Chapter 26
Are Aesthetics and Business Antagonists?

26.1 Commercial Aspects

We have talked about various elements to be thought about and considered in the construction of future music; now we ask: Where and how will I be positioned in the local or global marketplace? After all, you want or have to make a living out of your profession as a composer! Of course, you can become a university professor of composition to create financial stability, but we want to think about earning money as a composer, not as a teacher.

A composer's economic situation is complex and requires the ability to adapt while obtaining enough opportunities to stabilize a reputation for composing. Beethoven was moderately paid for the publication of his works, Charles Ives was an insurance salesman, and Cecil Taylor washed dishes. In examining many famous composers' economic struggles, it is clear that these prominent figures did whatever they had to do in order to survive financially while continuing to compose.

Commercially successful contemporary music composers, like Adi Yeshaya[1] (Fig. 26.1), see the marketplace through the following lens (refer to Chap. 8). He started in a time in which arranging music was equivalent to producing. This means that he uses his artistic sense to create but also had to consider what the public wanted and the commercial value of the work. As an orchestrator, even in an artistic kind of work, it has to satisfy a certain audience.[2] He acknowledges his advantageous position as important in creating financial stability but is also

[1]Adi Yeshaya is an Israeli-American composer, arranger and jazz pianist. Graduate of Berklee College of Music in Boston, Adi has been active, both as a musician and educator for nearly four decades. His credentials include arranging and orchestration for many artists, including Prince, Whitney Houston, Aretha Franklin, Lena Horne, Doc Severinsen, Burt Bacharach, The Minnesota Orchestra, and The Swiss Youth Jazz Orchestra/Jazzaar.

[2]Adi Yeshaya phone interview, March 31, 2019.

G. Mazzola et al., *The Future of Music*,
https://doi.org/10.1007/978-3-030-39709-8_26

Fig. 26.1 Composer Adi Yeshaya. Image reprinted with kind permission from Adi Yeshava

in tension with his artistic sensibility. He credits this balance as important to the construction of his identity as an artist.

The very topic of making money from composing is a difficult subject to many composers but, it is of course, a necessity it is worth spending some time thinking about why there is this discomfort. How can composers resolve a desire to create art for art's sake versus a need to construct commercially viable music? Is there a way to do both or are they always incompatible? In the following section, the rationale for why the very nature of being financially successful conflicts with artistic integrity will be explored. Further, possible changes in conceptualization about 'making it' will be proposed.

26.2 Conceptual Conflict of Money and Composition

26.2.1 Selling Beauty

Why is the commercial aspect of musical composition anathema to artistic sensibility? A quick answer would be that because music is about beauty, passion, and the highest spirituality—it is absurd to connect it to the down-to-earth question of making a living. Composers might prefer to live in the spirit of the Bible (Matthew 6:26): "Look at the birds of the air, that they do not sow, nor reap nor gather into barns, yet your heavenly Father feeds them."

However, this attitude is unrealistic. Composers do sow, so why shouldn't they gather? The serious relevance of composers' work in their communication with a community may establish a relationship of (mutual) appreciation. Why should this appreciation not be reflected in commercial value? But our question is somewhat more delicate: Is the aesthetic value of a musical work beyond any commercial valuation? And conversely: Is every commercial compensation of musical works their automatic degradation to merchandise? So let us envisage this provocative open question: Why shouldn't money be earned from selling beauty?

The context of commercial aspects of serious musical composition is immediately related to abusing the highest of human goods—in other words, as if it were "Love for Sale," to cite one of Cole Porter's jazz standards. The commercial system is in fact defined by an exchange of goods, of objects you can put onto a shelf to await purchase by a consumer. Music—if conceived in its full expressive power—it is a fully engaged dance of humans on the instrumental stage, which is incompatible with detachment and consumerism. How could such an intense existential utterance be deposited on a shelf like labeled consumer goods?

26.2.2 Artistic Integrity and the Pressure of Commercialization

Our context is also related to the gigantic music industry, where composers (called songwriters) are the first germ in a complex and brutal machinery between music publishing, performing, and recording. Although we understand every composer's disgust, a closer look at this environment is important. You have to know about copyright for music works (©) as opposed to music performances (℗). You have to know the copyright protection companies, otherwise you will be vulnerable whenever you sign a contract with one of those labels or publishing companies. In the music industry, composers, even if they enter with good intentions, will very probably be squeezed until all ideals are eliminated and Cole Porter's song becomes reality. Michael Jackson is a famous example of an artist who was exploited and deeply physically affected until his death. So the context is the music industry in which the environment resembles an industrial chimera—while the concern of a composer is to create living beauty. It could not be more split.

In this split world of commercial and aesthetic perspectives, there is one common concept that challenges all of them, and this is value. What is the value of what you are creating in a musical work? Is there any answer to the question of creating values here that might be valid in both systems? We suggest that our critical concept is this one: value.

26.2.3 Conflict of Art vs. Commercial

This split view makes the conflict complex: The artistic perspective is what the composer desires, the commercial one is what must be suffered. During Mao's cultural revolution, they killed intellectuals (anybody who was reading literature except Mao's red booklet was qualified as an intellectual) or sent them to the countryside to learn about "real life." It was an ideology that wanted to destroy the "high" bourgeois aspects of life—everything had to be low. A composer had to compose music for workers, not those "useless intellectual and decadent Western artifacts"; see the interview with Chinese composer Gen Gan-ru in [137] for more details.

The rigidity of this wall exists in the pair of opposites, "high" vs. "low". The question is whether we could resolve this tension without destroying one of them in the manner of Mao's cultural revolution.

26.3 Reconceptualizing to Allow for Monetary Value and Composition

26.3.1 Moving from High Versus Low to Inner Versus Outer

To begin with, the polarity "high" vs. "low" of life values is derived from an implicit gravitational force. Low is where you fall when you lose control over your body. High is where you jump when you are in full control. So what is that gravitation of life values, against which Goethe's Faust was fighting when the angels welcome him with the promise[3]: "Wer immer strebend sich bemüht, den koönnen wir erloösen."? It is the force of death against which every plant shows its triumph when growing towards the sky. But there is more than biology—it implicitly includes a divinity at which our high aspirations are targeted. There is a perfection and fulfillment toward which we are striving. For such a plan there is nothing more disturbing than everyday life's contingencies.

Perhaps we should spend a few minutes on divinities. After all, it's them who create the gravitational paradigm. The plan is to approach them and ideally be united. Every composer, when he/she creates that mini-universe of compositions, implicitly portrays a god's creational act. So let us question this attitude. Do we have to portray God when we compose? Or is there an alternative to this "divinity neurosis" that would not send us to Mao's farmers?

Our alternative in the present situation would then be to think about replacing the divinity paradigm with one where we do not need to 'jump to infinity' to achieve our compositional goals. Essentially, we could try to apply the same strategy as described in the instrumental context: embodiment instead of reference to divinities. Take the composition inside ourselves and fill it with the intensity of our lives. Becoming a composition with all our vital energies, being filled with our work as opposed to setting them down like a mini-universe creation. This being filled with our composition is a promising display: We have replaced the polarity of "high" vs. "low" with a new one: "inside" vs. "outside". We are identified with the composition, and our skin is the composition's skin, so to speak.

Why is this inside-outside paradigm a creative extension of the wall? Because now there is a natural interface between work and everyday life: the skin. There is an interface that before did not exist; there is no reconciliation between high and

[3]Whoever strives can be redeemed.

low. But the skin's paradigm offers an exchange of energies and ideas—its vibration can transmit messages from inside and also from outside. We are provided with a sensitive channel between the life of a work and the work of life.

26.3.2 Conceptualizing the 'Value' of Composition

It is about values when we confront the creation of musical beauty in a composition with the problem of making a living. In these values there is one common aspect, namely that both are about life: Composing is the innermost value of life and making a living is about life in a very basic sense. The wall that we have to face is the value for life. This wall unites the entire paradox of high artistic aspects of life and its low everyday aspect. We must break through this wall.

26.4 Final Step: Testing Our Extension

This metaphor is not only beautiful, it really means something in the understanding of one's compositional work and one's life environment. The work's pulsation can and should be felt in life, and the life's pulsation can and should be felt in the compositional development. Let us make a short example.

The Beatles, Bob Dylan, Witold Lutoslawski, and Iannis Xenakis modeled lives that balance the values of the aesthetic inside and living outside experiences without having to sacrifice one of the two sides of life. Their music is highly artistic and beautiful, and also became a part of our life.

Part VI
Commute

Chapter 27
CoMMuTE: Towards a Computational Musical Theory of Everything

Before we discuss the idea of a musical Theory of Everything (CoMMuTE), we recall the physical ToE and its motivations.

27.1 The Physical Theory of Everything (ToE)

A physical Theory of Everything claims the integration of the four fundamental physical force types: electromagnetic, weak, strong, and gravitation. The electromagnetic and weak are already united, and called "electro-weak". The still hypothetical integration of the electro-weak force with the strong force is called GUT: Grand Unification Theory. Integration means that all these forces are special cases of a fundamental force, which splits into the four forces by a breaking of structural symmetries when energies are below a threshold.

Let us stress that already the reduction to those four forces is anything but evident. For example, would you guess that the mechanical forces of, say, a hammer hitting a nail, are the same type as the adhesion forces of glue? Or the forces of chemical reactions? Or the force of sunlight tanning your skin? They all pertain to the electromagnetic force type. The physical sciences and their outlets in chemistry have achieved an incredible reduction of the apparent variety of force types. This means that the surface of physical actions does not prevent a deep theory from unifying superficial diversities. When we will discuss the situation in music, we should keep in mind this fact.

It is remarkable that "Everything" in physics relates to physical forces, but not to psychological or symbolic realities. This restriction is significant since it is different from any physicalist totalitarianism. Physics does not claim a total explanation of this world: physics deals with the outer nature and has never tried to reduce psychological or symbolic realities to physics. Instead, prominent theoretical physicists, such as Roger Penrose, see Fig. 27.1, argue that the innermost physical

G. Mazzola et al., *The Future of Music*,
https://doi.org/10.1007/978-3-030-39709-8_27

Fig. 27.1 Theoretical Physicist Roger Penrose. © Photo Researchers/Science History Images/Alamy

ontology might rely on mathematics. He also argues in [138] that physics has not yet included the psychological realm of the human mind in its basic conceptual architecture.

Despite the success of simplifying physical force types to three (electro-weak, strong, and gravitation), it is not clear why the ToE and even the GUT should work. Nevertheless, physicists seem to believe in an ultimate unification. Some might adhere to a monotheist paradigm: There is only one innermost—well yes: divine—entity that shapes the universe. The success of physics gives them enough motivation to work in this direction with an impressive shared social, organizational,[1] and economic effort called "Big Science"; see our discussion of Big Science in Chap. 9.

27.2 Why Would We Think About a Musical ToE (CoMMUTE)?

In view of the fundamental role of music in all cultures[2] which we have discussed in Chap. 3, it is not astonishing to think of a unified view of music in the spirit of the physical ToE:

> Could it be that the variety of musical expressivity is the unfolding of a unique fundamental "force field"?

Before we delve into this hypothesis of musical unification, we should first understand the wording CoMMUTE. It means *Computational Music Theory of Everything*, but it would not, similar to ToE, include strictly everything. This is

[1]The Internet was invented by Tim Berners Lee at the CERN to coordinate nuclear research efforts globally.

[2]Even where music is virtually forbidden, with the Taliban, for example, its force is recognized, and that is why it is forbidden, sad irony.

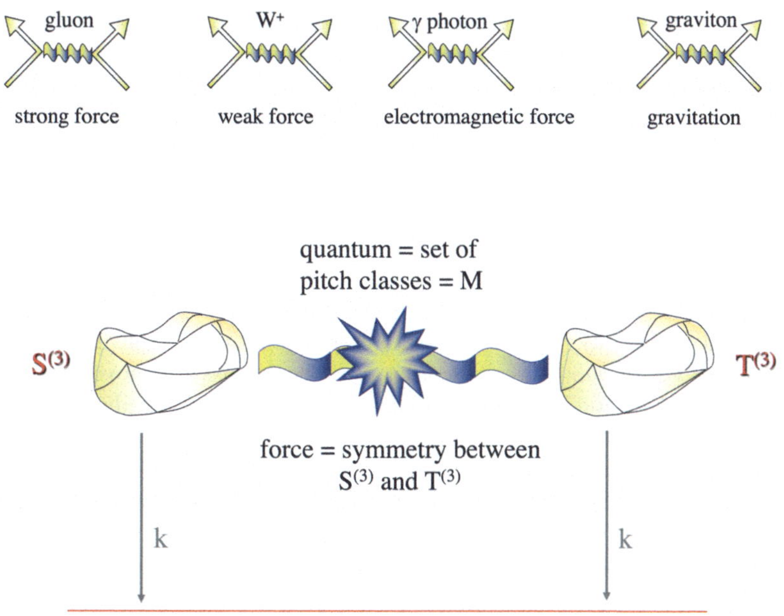

Fig. 27.2 Physical quanta and musical quantum M for a modulation from tonality S to tonality T; k is the cadence. © Guerino Mazzola

also why we add the adjective "computational". The idea is to think of a music theory that is computational; other theories might exist, but they are not addressed here.

Moreover, it is also, similar to ToE, not intended to subsume all realities, the physical, psychological, or symbolic (introduced in Chap. 2). Nevertheless, recall from Chap. 4 that the musical idea has been a driving force in the development of physics and astronomy from Pythagoras to String Theory. It is not evident in how far a COMMUTE would connect to ToE, but one should keep in mind these deep relationships between music and physics, especially when composers wants to justify their overall philosophical motivation.

For COMMUTE the analogy to the physical ToE is problematic, since ToE is a precise hypothesis, the unification of all physical force types, independently of how this would work. Music does not share the simple idea of forces which are embodied by quanta.[3] Nevertheless, the core idea of Mazzola's modulation model (see also Chap. 12 and Fig. 27.2) was an exact analogy to physical force quanta, musical symmetries being the analogy of forces, while modulation quanta (sets of pitches) are the analogy of physical quanta. But this is a special situation that cannot for the time being be extended to modulations among non-diatonic scales.

[3] These are photons for the electromagnetic force, W and Z bosons for the weak force, gluons for strong force, and hypothetical gravitons for gravitation.

Another type of forces was used in the mathematical model of counterpoint (see Sect. 15.1). Here, the dichotomy of consonances and dissonances was deformed by a symmetry in the role of an elastic force acting on the set of intervals.

The idea of forces in music (theory) is not new, see [100]. Already Schoenberg used the metaphor of erotic (!) forces to explain harmonic tension in [146]. But there is no such universally acclaimed theoretical architecture as in physics, forces may show up (metaphorically or literally), but music has a number of very different structural paradigms, such as local-global duality, geometric ideas (the harmonic Moebius strip, for example, see Sect. 12.4), or gestural and topological approaches.

What are arguments against such a strong CoMMuTE hypothesis? To begin with, the individual creativity of a musical composer or improviser seems to forbid any "universalist" background. This is the heritage of the Renaissance movement, which opposes to the Pythagorean "world formula" in an irreducible individual genealogy. This logical opposition is however not stringent since a painter, for example, may create a deeply individual work of art with colors being completely described by the electromagnetic force that defines light and its action within the human eye. And on a higher level of structural abstraction, the variety of musical transformations can be described by a huge mathematical group, typically ranging within the cardinality[4] of 10^{40}. This is all comprised within a very clear and unified conceptual architecture of theory, but such a virtually infinite number guarantees an unlimited variety of individual utterances.

Another argument against CoMMuTE would be the suspicion that this hypothesis is a consequence of the Western (Christian) colonialist mentality. Are we trying to unite all musical cultures under a big equalizing umbrella? And thereby destroying unsurmountable differences? This is a delicate question since we already have the example of a destructive reduction when transcribing Arab Maqam music to Western notation—all the essential pitch bend effects are eliminated. Other examples of the same type are abundant. But such a translational pathology may be eliminated by a more diligent conceptualization. For example, the language of denotators as described in Sect. 10.1.1 can describe musical objects much better than the traditional Western score. This means that this question could be answered by extending the given language to a state where some differences would be taken care of.

It is a delicate question because the argument of an extended language seems to be purely formal, it would not touch the cultural differences of the role of music. For example, the African social role of music is radically different from the Western role, and also different from the Indian role of Raga music, say. This aspect relates

[4]The number of transformations on the local score of the *presto*® software described in Sect. 17.2.2 is

$$10'445'260'466'832'483'579'436'191'905'936'640'000 \approx 1.04453 \times 10^{37}.$$

to music sociology or psychology. We do not include these aspects in the hypothesis of ComMute. At the time being such an inclusion seems too ambitious and also dangerous for the named reasons. But we shall see below in Sect. 27.3.3 that on the level of music theory, relations between European counterpoint and Raga music are appearing. Perhaps one should also reflect upon the idea of "fundamentally incomparable cultures" in view of a separation and even the famous "clash of civilizations". On the level of human rights the vast majority of cultures have agreed on a shared canon, such as the *habeas corpus* principle. In the movie *Teak Leaves at the Temples* [132], Mazzola argued that the language of gestures might be an approach to a non-divisional understanding of musical cultures.

Let us complete this introduction with a discussion of the specification "computational" in ComMute. This means two things. First, such a music theory should be covering all that is accessible by mathematical methods and concepts. Second, it should be accessible via computational engines such as computers. This is a restriction of "everything" to "computational everything". It has the same function as in physics: What is not accessible in this way is not physically relevant. It is simply an expression of modesty, one only considers topics that are 'visible' to computation. Of course, as musical research progresses, more and more things may become 'visible'. But it is also risky to restrict one's views to what is actually computable. For example, Stephan Hawking claimed that the concept of a God is superfluous in physics. He did not consider the language and spirit of mathematics as being a *conditio sine qua non* for physics. The computational principle is pre-physical, and Hawking forgot to ask where we get that language from. Radical neuroscientists might argue that mathematics is an artifact of neurons, i.e., a product of physical reality. But any proof thereof would use huge mathematical tools, which creates a circular argument, it would not explain anything.

In this sense, ComMute is a methodological limitation and should not be taken as a definition of music, but as a perspective, which can be tested and used to shape concrete progress. Nevertheless, the hypothesis is a very strong one, similar to ToE, or even more tricky because the conceptual landscape in music is less unified than physics.

27.3 Some Directions Towards ComMute

In the following sections we want to describe a number of vectors towards such a ComMute, without claiming completeness.

27.3.1 Harmony and Rhythm

Harmony and rhythm have played very different roles in the history and cultural diversity of music. For example, recent research [129] stresses the fundamental difference in harmony and rhythm between classical European and African music. The complexity of Western harmony corresponds to the complexity of African polyrhythms, and vice versa: the simplicity of Western rhythms corresponds to the simplicity of African harmony. This difference however does not mean that rhythm cannot be dealt with like harmony. In fact, both phenomena deal with periodic sets of events, octave periodicity in harmony vs. time periodicity of rhythm. In MaMuTh, Mazzola's modulation model is not limited to pitch, but can equally be applied to time, just rotate pitch by 90° into time. This has been used for rhythmic modulation in the first movement of the composition *Synthesis* discussed in Sect. 17.2.3.

This double periodicity has also been used to investigate periodic structures in pitch or time using the finite Fourier decomposition of periodic functions. The classical theory of Fourier for time functions is a classical theory (partials, Fast Fourier Transform, etc.), but the (octave-)periodic pitch functions have been analyzed only recently by David Lewin, Ian Quinn, and Emmanuel Amiot [9]. This research proves that harmony and rhythm could converge to a unified theory of periodic functions in a number of parameter spaces. And such a unification could eventually create a less diversified perspective of Western and African musical cultures. Of course, the difference between music sociology of Europe and Africa remains untouched: The African music culture is shared by everybody, from childhood to adult life, Africans are musicians, the European musician as a specifically educated person is not a standard role in Africa. This difference may vanish in the future if we learn to think in pitch and time according to the same theoretical and compositional paradigms.

27.3.2 Gestures for Harmony and Counterpoint

In MaMuTh, modulation theory and counterpoint were developed using symmetries on pitch class spaces. This was not a topological approach and had its limits. For modulation theory the paring of tonalities was only permitted for tonalities in the same orbit of a symmetry group. Modulation from a major tonality to a gipsy tonality or even a pentatonic one was not conceived. A similar restriction happened to contrapuntal concepts. This model used a finite number of consonances selected by specific symmetries. It would crash if we had to consider an infinity of consonances in a continuous and therefore infinite interval model of counterpoint.

Both restrictions could be solved by a shared new language: musical gestures. Modulation could be remodeled as a gestural deformation of chords instead of a symmetric action [122]. Such a deformation does no longer require the two tonalities living in the same symmetry orbit. Using gestures and their homology theory, counterpoint could be remodeled independently of the number of consonant intervals at stake [7, Ch. 10].

This is an example of gesture theory being a unifier of music theory, an interesting parallel to physics, where String Theory is a strong candidate for ToE. Strings are analogues to gestures in music theory, in other words: gestures are the musical analog to strings.

27.3.3 Counterpoint Worlds for Different Musical Cultures

As discussed in Sect. 15.1.3.2, the mathematical theory of counterpoint has not only embraced Fux tradition, but also opened with its five new worlds connections to Raga music and Scriabin's mystic chord. This theory is on its way to a global counterpoint theory. Connecting counterpoint models with Indian music is a sensational bridge between totally different musical cultures. We should recall here that connections between Western and Raga music have been investigated by Robert Morris and Chitravina N. Ravikiran under the title of "Melharmony" [140]. This theory aims to create chords and counterpoints based on the melodic rules of evolved systems across the world. It could happen that our counterpoint worlds and melharmony converge to a new synthesis of two strong theoretical traditions.

27.3.4 Complex Time for Unification of Mental and Physical Realities in Music

We have discussed complex time in Chap. 13. This concept unifies Descartes' *res extensa*, the physical reality, with *res cogitans*, the thinking, mental reality. This is a strong step towards a unification of a well-known duality in music: thinking and making, the physical utterance of performance vs. the mental construction in composition, performance, and improvisation. In that chapter, we have sketched ways of structurally connecting these two realities by means of world sheets, which are completely analogous to world sheets in physical String Theory.

The duality of thinking and making in music has been a dividing force between theorists and performers in the Western music world. It is also present in the academic structure where "applied" scholars (teaching instrumental fields) are separated from "academic" scholars (music theory, musicology, education, ethnomusicology, music psychology). Their interaction is reduced to a poor "laissez vivre" and is not based on a shared reality, a fact that complex time theories might annihilate in the future.

27.3.5 Symbolic and Real Gestures

The idea of complex time has also been applied to connect symbolic gestures on the score to real gestures of the performing artist, see Sect. 13.5. This result fosters the unification of thinking and making. Eventually, this should help unify these two aspects of the art of music to a comprising reality.

27.3.6 Unifying Note Performance and Gestural Performance: Lie Operators

Computational performance theory was strongly developed by the Stockholm group around Johan Sundberg [35]. This approach was concerned with the transformation of notes to sound events. It could be shown by the work of the Zurich group of Mazzola [114, Ch. 39.7] that important cases of such a performative transformation can be described by classical Lie operators from differential geometry. More precisely, the transition from a given performance stage to a more refined one is described by a Lie operator acting on the given performance vector field. When performance theory was extended to gestural performance, i.e. the transformation of symbolic gestures in the score to physical gestures of a musician, it could be shown that the same Lie operator formalism can be carried over to gestures [113, Ch. 78.2.13]. In other words, *we now have a unified formalism of performance theory for notes and for gestures.*

27.3.7 Unifying Composition and Improvisation?

It is an open question in how far composition and improvisation could be unified as special cases of a unique still hidden dynamics. We know of many famous composers, such as Beethoven or Mozart, that their compositions were often created from improvisation, see the genesis of Beethoven's *Sonata Op. 109* [96], for example. In the Indian tradition of Raga, improvisation and composition are intimately related on the basis of mela scales (recall Sect. 15.1.3.2). Perhaps the combination of flow concepts and gestures, as sketched in the free jazz book [116], could help find a unified understanding of musical creativity in composition and improvisation.

27.4 Imagining Big Science for ComMute

A future music theory that includes Big Science should above all change some of the dominating traditional characteristics of music theory and musicology as follows:

1. *Change "Meditative" into "Operational"!* We argue that it is wiser to try out several variants of a composition on a sequencer than to meditate on a fictive best solution.
2. *Change "Metaphoric" into "Explicit"!* For example, it is advantageous to look for tempo curves instead of nebulous metaphors of movement.
3. *Change "Ubiquity" into "Topography"!* In fact, it is nonsense to postulate that the musical work has a monolithic omnipresence—and to be deceived if you won't find it that way round. Better look for topographically differentiated traces of what could contribute to a distributed concept of the musical work.
4. *Change "Ontological" into "Semiotic"!* Musicology often tries to find "the true and essential meaning" of something instead of distinguishing between layers of signification which are best described by semiotic categories such as denotation or connotation.
5. *Change "Magic" into "Communicative"!* The old-fashioned and absurd swearing to the genius of a great composer should be replaced by communication of ideas and perspectives.
6. *Change "Transcendence" into "Precision"!* It would be more interesting and scientific to publish detailed analyses of musical works instead of writing all those feuilletonistic books on composers. After all, the tools are ready; the minds should realize this.

This list might sound provocative, but without these changes no Big Science in music will be feasible. Perhaps are we now facing a transformation of music theory that is comparable to the Galilean transformation, which replaced reading Aristotelian philosophical books on physics by computational and experimental methods.

With this in mind, musical creativity will dramatically change its face with Big Science. With the elimination of the outdated genius paradigm, music will be more of a collaborative research with huge technological tools for its globally distributed realization. This collaborative style will also eliminate the strict separation between artist and audience. And the artistic virtuosity in the performance of musical works will become a niche exercise that can be performed better by robots. This does not mean that human instrumental virtuosity will disappear, but it will play the completely different role of a thinking-making research, not of a (however sophisticated) performance of given templates.

27.5 Hegel's Weltgeist and the Big Bang

To conclude this prospective final chapter, we would like to observe a fascinating parallel between the physical world and the philosophical realm of thoughts. In physics, we have learned that the universe is an evolving entity, starting some 13.8 billion years ago with the Big Bang and evolving through the early generation of physical laws (during the Planck time of 5.39×10^{-44} s), the separation of the four fundamental forces, and then the creation of elementary particles, the atoms, the stars, and the solar system with its rich biochemistry that gave rise to the human race and its culture. It is remarkable that philosopher Georg Wilhelm Friedrich Hegel in his idealistic system also conceives an evolution, but this time a spiritual one, which brings his Weltgeist (world spirit) to consciousness, his Weltgeist is an awaking entity, which successively unfolds more sophisticated structures and capacities. It is more than an isolated entity, it is the world's mental substance. Confronting the Big Bang with the Weltgeist, one could argue that these are parallel theories, and that there could be a common ground for their reality. In other words: is the Weltgeist a Big Bang of Descartes' *res cogitans*, while the Big Bang would be the awakening of Descartes' *res extensa*? And could this parallelism be the surfaces of a common entity, which is an awaking universe of everything, of thoughts and things?

It is interesting to view the musical realm as a model for such a total unification of thinking and making, of the passion of existence: the synthesis of passive and active.

References

1. T.W. Adorno, *Zu einer Theorie der musikalischen Reproduktion* (Suhrkamp, Frankfurt/M, Berlin, 1946)
2. T.W. Adorno, *Fragment über Musik und Sprache* (J.B. Metzler, Stuttgart, 1956)
3. T.W. Adorno, *Towards a Theory of Musical Reproduction: Notes, a Draft and Two Schemata*, 1st edn. Polity (Wiley, London, 2006)
4. K. Agawu, *African Rhythm* (Cambridge University Press, Cambridge, 1995)
5. K. Agawu, *Representing African Music* (Routledge, New York, 2003)
6. O.A. Agustín-Aquino, G. Mazzola, Contrapuntal aspects of the mystic Chord and Scriabin's Piano Sonata No. 5, in *Mathematics and Computation in Music* (Springer, Cham, 2019), pp. 3–20
7. O.A. Agustín-Aquino, J. Junod, G. Mazzola, *Computational Counterpoint Worlds*. Springer Series Computational Music Science (Springer, Heidelberg, 2015)
8. R. Alpaydin, G. Mazzola, HarmoRUBETTE reconsidered. Musicae Scientiae (2015)
9. E. Amiot, *Music Through Fourier Space*. Springer Series Computational Music Science (Springer, Heidelberg, 2016)
10. J.J. Arnett, Adolescents' uses of media for self-socialisation. J. Youth Adolesc. **24**, 519–533 (1995)
11. I. Bars, J. Terning, *Extra Dimensions in Space and Time* (Springer, Berlin, 2010)
12. W. Benjamin, *The Work of Art in the Age of Mechanical Reproduction* (Schocken Books, New York, 1936), p. 21
13. D. Boer, *Music Makes the People Come Together: Social Functions of Music Listening for Young People Across Cultures* (Department of Psychology, Victoria University of Wellington, Wellington, 2009). http://researcharchive.vuw.ac.nz/bitstream/handle/10063/1155/thesis.pdf?sequence=1
14. A. Boissière, Geste, interprétation, invention selon Pierre Boulez. Revue DEMéter, déc (2002), Univ. Lille-3
15. P. Boulez, *Structures, Premier Livre* (UE, London, 1953)
16. P. Boulez, Musikdenken heute I,II; Darmstädter Beiträge V, VI. Schott, Mainz 1963 (1985)
17. J. Brahms, Klaviersonate C-Dur Op. 1. Wiener Urtext Edition (2009)
18. E. Bullough, Recent work in experimental aesthetics. Br. J. Psychol. **12**, 76–99 (1921)
19. C. Buteau, Motivic Spaces of scores through RUBATO's MeloTopRUBETTE, in *Perspectives in Mathematical and Computational Music Theory*, ed. by G. Mazzola, T. Noll, E. Lluis-Puebla (Verlag epOs-Music, Osnabrück, 2004), pp. 330–342
20. T. Campanella, *Philosophiae rationalis partes quinque* (Apud J. et T. Du Bray, Paris, 1637–1638)

G. Mazzola et al., *The Future of Music*,
https://doi.org/10.1007/978-3-030-39709-8

21. J. Cavaillès, *Méthode axiomatique et formalisme* (Hermann, Paris, 1938)
22. T. Chamorro-Premuzic, A. Furnham, Personality and music: can traits explain how people use music in everyday life. Br. J. Psychol. **98**, 175–85 (2007). https://doi.org/10.1348/000712606X111177
23. E. Chew, A. Volk, C.-Y. Lee, Dance music classification using inner metric analysis, in *Proceedings of the 9th INFORMS Computer Society Conference* (2005), pp. 355–370
24. G. Cybenko, Complexity theory of neural networks and classification problems, in *Neural Networks*, ed. by L.B. Almeida, C.J. Wellekens, vol. 412 (Springer, Berlin, 1990)
25. M. Davis, Bitches Brew. 2 LPs, Columbia (1969)
26. R. Descartes, *Principia Philosophiae (Principles of Philosophy)*. Translation with explanatory notes by Valentine Rodger and Reese P. Miller (Reprint ed.) (Reidel, Dordrecht, 1644)
27. "DJ Q Bert 1991 DMC Championship," uploaded by Boom Bap Focus (2017), https://www.youtube.com/watch?v=8wITk8-7oNU
28. S. Feld, *Jazz Cosmopolitanism in Accra* (Duke University Press, Durham, 2012)
29. A. Fleischer, *Die analytische Interpretation* (dissertation.de, Berlin, 2002)
30. A. Fleischer, Metric investigations in Brahms' Symphonies, in *Perspectives of Mathematical and Computational Music Theory*, ed. by G. Mazzola, T. Noll, E. Lluis-Puebla (epos music, Osnabrück, 2004)
31. D. Falk, Prelinguistic evolution in early hominins: whence motherese. Behav. Brain Sci. **27**, 491–503 (2004). https://doi.org/10.1017/S0140525X04000111
32. D. Falk, The "putting the baby down" hypothesis: bipedalism, babbling, and baby siblings. Behav. Brain Sci. **27**, 526–534 (2004). https://doi.org/10.1017/S0140525X0448011X
33. A. Fleischer, T. Noll, Analytical coherence and performance regulation. J. New Music Res. **15**, 314–325 (2002)
34. C. Fourier, L'Harmonie universelle et le phalanstère, tomes 1 et 2. (1849) Hachette Livre, Paris (2018)
35. A. Friberg, R. Bresin, J. Sundberg, *Overview of the KTH Rule System for Music Performance*. Advances in Experimental Psychology, special issue on Music Performance (2006)
36. H. Fripertinger, Endliche Gruppenaktionen in Funktionenmengen - Das Lemma von Burnside - Repräsentantenkonstruktionen - Anwendungen in der Musiktheorie. Dissertation Univ. Graz (1993)
37. W. Frisch, *Brahms and the Principle of Developing Variation*. No. 2 (California Press, California, 1990)
38. J.J. Fux, Gradus ad Parnassum (1725). Dt. und kommentiert von L. Mitzler, Leipzig 1742. English edition: The Study of Counterpoint. Translated and edited by A Mann (Norton & Company, New York, 1971)
39. P. Gilroy, *The Black Atlantic: Modernity and Double Consciousness* (Verso, New York, 1993)
40. G. Gould, *Beethoven's Op. 57. The Glenn Gould Collection*, vol. 1 (Sony, 2012)
41. W. Graeser, Bachs "Kunst der Fuge", in *Bach-Jahrbuch* (1924)
42. E. Hanslick, Vom Musikalisch-Schönen. Breitkopf und Härtel (1854), Wiesbaden (1980)
43. S. Hawking, *A Brief History of Time* (Bantham Books, New York, 1988)
44. M.H. Hayes, "The writing on the wall," taken by surprise, in *Connecticut*, ed. by A.C. Albright, D. Gere (Wesleyan University Press, Middletown, 2003), p. 108
45. T. Hays, V. Minichiello, The meaning of music in the lives of older people: a qualitative study. Psychol. Music **33**, 437–51 (2005). https://doi.org/10.1177/0305735605056160
46. S. Hollos, R.I. Hollos, *Creating Rhythms* (Exstrom Laboratories LLC, Longmont, 2014)
47. S. Hollos, R.I. Hollos, *Creating Melodies* (Exstrom Laboratories LLC, Longmont, 2018)
48. https://www.ableton.com/en/push
49. https://www.crunchbase.com/organization/acpad-instruments
50. https://www.risingsoftware.com/auralia
51. https://www.pgmusic.com/

52. https://ccrma.stanford.edu/software/stk
53. http://www.cern.ch
54. http://chuck.cs.princeton.edu
55. https://www.cme-pro.com/xkey-air
56. https://www.ars-nova.com/cp
57. https://csound.com
58. https://new.steinberg.net/de/cubase
59. https://new.steinberg.net/de/dorico
60. http://endel.io/about
61. https://extemporelang.github.io
62. https://www.akaipro.com/ewi-usb
63. https://www.finalemusic.com
64. https://www.image-line.com/flstudio
65. http://www.pawfal.org/fluxus
66. https://charlie-roberts.com/gibber
67. https://www.cs.hmc.edu/~keller/jazz/improvisor
68. http://www.ixi-audio.net/content/software.html
69. https://docs.oracle.com/javase/tutorial/java/concepts/index.html
70. https://docs.oracle.com/javase/tutorial/java/concepts/object.html
71. https://docs.oracle.com/javase/tutorial/java/concepts/class.html
72. https://www.apple.com/logic-pro
73. https://cycling74.com
74. http://en.wikipedia.org/wiki/Celemony#Melodyne
75. http://www.midi.org
76. https://www.native-instruments.com/de/products/maschine/production-systems/maschine
77. http://nodalmusic.com
78. https://www.noteperformer.com
79. https://medium.freecodecamp.org/object-oriented-programming-concepts-21bb035f7260
80. http://opensoundcontrol.org/introduction-osc
81. http://overtone.github.io
82. https://www.avid.com/pro-tools
83. https://puredata.info
84. https://www.reaper.fm
85. https://www.roland.com/us/products/fa-08
86. https://www.avid.com/sibelius-ultimate
87. https://sonic-pi.net
88. https://supercollider.github.io
89. https://www.syntorial.com
90. https://tidalcycles.org
91. http://www.timothywisdom.com/djtaxonomy/scratching.html
92. https://www.researchgate.net/figure/A-scratch-composition-sketched-in-a-custom-notation-format-with-notated-rhythm_fig4_233814981
93. https://soundcloud.com/vocaloid_yamaha
94. https://www.youtube.com/watch?v=VOat_xsGUOw
95. R. Jackendoff, F. Lerdahl, *A Generative Theory of Tonal Music* (MIT Press, Cambridge, 1983)
96. W. Kinderman, *Artaria 195* (University of Illinois Press, Urbana, 2003)
97. D.L. Kohut, *Musical Performance: Learning Theory and Pedagogy* (Stipes Publishing LLC, Champaign, 1992)
98. K. Komparu, *The Noh Theater* (Waterhill, New York, 1983)
99. J.D. Kramer, *The Time of Music* (Schirmer, New York, 1988)
100. S. Larson, *Musical Forces* (Indiana U Press, Bloomington, 2012)
101. A. Lepecki, Choreopolice and Choreopolitics: or, the task of the dancer. Drama Rev. **57**(4), 16 (2013)

102. G. Ligeti, Pierre Boulez: Entscheidung und Automatik in der Structure Ia. Die Reihe IV, UE, Wien (1958)
103. A.J. Lonsdale, A.C. North, Why do we listen to music: a uses and gratifications analysis. Br. J. Psychol. **102**, 108–34 (2011). https://doi.org/10.1348/000712610X506831
104. P.D. MacLean, The triune brain, emotion, and scientific bias, in *The Neurosciences: Second Study Program*, ed. by F.O. Schmitt (Rockefeller University Press, New York, 1970), pp. 336–48
105. M. Mannone, G. Mazzola, Hypergestures in complex time: creative performance between symbolic and physical reality, in *Springer proceedings of the MCM15 Conference* (2015)
106. M. Maurizio Gentilucci, M.C. Corballis, From manual gesture to speech: a gradual transition. Neurosci. Biobehav. Rev. **30**(7), 949–960 (2006)
107. G. Mazzola, *Gruppen und Kategorien in der Musik* (Heldermann, Berlin, 1985)
108. G. Mazzola, *Geometrie der Töne* (Birkhäuser, Basel, 1990)
109. G. Mazzola, *Synthesis*. SToA 1001.90, Zürich (1990)
110. G. Mazzola, L'essence du bleu Acanthus (2002)
111. G. Mazzola, *Musical Performance—A Comprehensive Approach: Theory, Analytical Tools, and Case Studies*. Springer Series Computational Music Science (Springer, Heidelberg, 2010)
112. G. Mazzola, Mathematical music theory and the musical math game – two creative ontological switches, in *Proceedings of the MCM Conference 2017, UNAM Mexico* (Springer, Heidelberg, 2018)
113. G. Mazzola, *The Topos of Music III: Gestures*. Springer Series Computational Music Science (Springer, Heidelberg, 2018)
114. G. Mazzola, *The Topos of Music II: Performance*. Springer Series Computational Music Science (Springer, Heidelberg, 2018)
115. G. Mazzola, M. Andreatta, Diagrams, gestures, and formulas in music. J. Math. Music **1**(1), 23–46 (2007)
116. G. Mazzola, P.B. Cherlin, *Flow, Gesture, and Spaces in Free Jazz—Towards a Theory of Collaboration*. Springer Series Computational Music Science (Springer, Heidelberg, 2009)
117. G. Mazzola, J. Park, Passionate Message. Silkheart Records, SHCD159 (2012)
118. G. Mazzola, et al., *The Topos of Music—Geometric Logic of Concepts, Theory, and Performance* (Birkhäuser, Basel, 2002)
119. G. Mazzola, J. Park, F. Thalmann, *Musical Creativity* (Springer, Heidelberg, 2011)
120. G. Mazzola, M. Mannone, Y. Pang, *Cool Math for Hot Music* (Springer, Heidelberg, 2016)
121. G. Mazzola, M. Mannone, M. O'Brien, Y. Pang, N. Torunsky, *All About Music* (Springer, Heidelberg, 2016)
122. G. Mazzola, Gestural dynamics in modulation: (towards) a musical string theory, in *The Musical-Mathematical Mind*, ed. by G. Pareyon, S. Pina-Romero, O.A. Agustin-Aquino, E. Lluis-Puebla. Springer Series Computational Music Science (Springer, Heidelberg, 2017)
123. M. Merleau-Ponty, Phénoménologie de la perception. Gallimard (1945)
124. G. Miller, Evolution of human music through sexual selection, in *The Origins of Music*, ed. by N.L. Wallin, B. Merker, S. Brown (The MIT Press, Cambridge 2015), pp. 329–360
125. G. Milmeister, *The Rubato Composer Music Software: Component-Based Implementation of a Functorial Concept Architecture*. Springer Series Computational Music Science (Springer, Heidelberg, 2009)
126. H. Minkowski, *Space and Time: Minkowski's Papers on Relativity* (Minkowski Institute Press, Montreal, 2012)
127. R. Moog, MIDI, musical instrument digital interface. J. Audio Eng. Soc. **34**(5), 394–404 (1986)
128. F. Moten, *In the Break: The Aesthetics of the Black Radical Tradition* (University of Minnesota Press, Minneapolis, 2003), p. 39
129. G. Munyaradzi, W. Zimidzi, Comparison of western music and African music. Creat. Educ. **3**(2), 193–195 (2012)
130. J.-J. Nattiez, Fondements d'une Sémiologie de la Musique. Edition 10/18 Paris (1975)

131. J. Noer "J-Sun", *Music, Moves, and Mind-Sets: A B-Boy's Guide to Applying Hip Hop Pedagogy in Education and Hip Hop Resource Archive* (Kendall Hunt Publishing, Iowa, 2013)
132. G. Nugroho, Teak Leaves at the Temples. DVD, Trimax Enterprises Film (2007)
133. P. Oliver, *Savannah Syncopators* (Studio Vista, London, 1970)
134. B. Owsinski, *Music 4.1* (Hal Leonard Books, Milwaukee, 2016)
135. Y. Pang, Rural and Urban Explorations of Transnational Performance: A Dialogue Between Traditional Southwestern Chinese and Contemporary North American Music & Dance (2018). Retrieved from the University of Minnesota Digital Conservancy. http://hdl.handle.net/11299/202159
136. J. Panksepp, G. Bernatzky, Emotional sounds and the brain: the neuro-affective foundations of musical appreciation. Behav. Process. **60**, 133–55 (2002). https://doi.org/10.1016/S0376-635700080-3
137. W. Parker, Conversations. Rogueart (2011)
138. R. Penrose, *The Road to Reality* (Vintage, London, 2002)
139. R. Pfeifer, J. Bongard, *How the Body Shapes the Way We Think: A New View of Intelligence* (MIT Press, Cambridge, 2006)
140. C.N. Ravikiran, Robert Morris and the concept of Melharmony. Perspect. New Music **52**(2), 154–161 (2014)
141. B. Riemann, Über die Hypothesen, welche der Geometrie zugrunde liegen (Habilitationsreferat 1854). Gott. Abh. No. 13 (1867)
142. H. Riemann, Vereinfachte Harmonielehre oder die Lehre von den tonalen Funktionen der Akkorde. London (1893)
143. T. Schäfer, P. Sedlmeier, C. Städtler, D. Huron, The psychological functions of music listening. Front. Psychol. **4**, 511 (2013)
144. W. Schmidt-Biggemann, *Topica universalis* (Meiner, Hamburg, 1983)
145. J.-C. Schmitt, La raison des gestes dans l'Occident médiéval. Gallimard, Paris (1990)
146. A. Schönberg, Harmonielehre (1911). Universal Edition, Wien (1966)
147. A. Schönberg, Brahms der Fortschrittliche, in *Stil und Gedanke* (1933/47) (Fischer, Frankfurt am Main, 1995)
148. W.-A. Schultz, Die Heilung des verlorenen Ichs - Kunst und Musik in Europa im 21. Jahrhundert (Europaverlag, München, 2018)
149. R. Sessions, *Questions About Music* (W. W. Norton and Company, New York, 1971)
150. S. Silver, Dance videos. https://www.sherriesilver.org
151. L.R. Squire, N. Butters, *Neuropsychology of Memory* (Guildford, New York, 1984)
152. Tupac. https://de.wikipedia.org/wiki/Tupac_Shakur
153. P. Valéry, in *Cahiers I–IV (1894–1914)*, ed. by N. Celeyrette-Pietri, J. Robinson-Valéry (Gallimard, Paris, 1987)
154. A. Valle, *Introduction to SuperCollider* (Logos Verlag Berlin GmbH, Berlin, 2016)
155. B. Weinstein, "The Collaborative Legacy of Merce Cunningham," Places Journal (2011). https://placesjournal.org/article/the-collaborative-legacy-of-merce-cunningham
156. R. Wendorff, *Zeit und Kultur* (Westdeutscher Verlag, Opladen, 1980)
157. R. Wieland, J. Uhde, *Forschendes Üben* (Bärenreiter-Verlag, Kassel, 2002)
158. J. Winson, *Brain and Psyche. The Biology of the Unconscious* (Anchor Press/Doubleday, Garden City, 1985)
159. N. Yoneda, On the homology theory of modules. J. Fac. Sci. Univ. Tokyo **7**, 193–227 (1954)

Index

G. Mazzola et al., *The Future of Music*,
https://doi.org/10.1007/978-3-030-39709-8

The manufacturer's authorised representative in the EU is Springer Nature Customer Service Centre GmbH, Europaplatz 3, 69115 Heidelberg, Germany. If you have any concerns regarding our products, please contact ProductSafety@springernature.com

Printed and bound by CPI Group (UK) Ltd, Croydon, CR0 4YY
30/09/2026
02225647-0003